$2.00

G. F. GAUSE

The Struggle for Existence

A Classic of
Mathematical Biology
and Ecology

DOVER BOOKS ON
THE BIOLOGICAL SCIENCES

(continued on back flap)

THE STRUGGLE
FOR
EXISTENCE

THE STRUGGLE
FOR
EXISTENCE

BY

G. F. GAUSE
Zoological Institute of the University of Moscow

Dover Publications, Inc.
New York

This Dover edition, first published in 1971, is an unabridged and unaltered republication of the work originally published by The Williams & Wilkins Company in 1934.

International Standard Book Number: 0-486-22697-2
Library of Congress Catalog Card Number: 71-143675

Manufactured in the United States of America
Dover Publications, Inc.
180 Varick Street
New York, N. Y. 10014

FOREWORD

For three-quarters of a century past more has been written about natural selection and the struggle for existence that underlies the selective process, than perhaps about any other single idea in the whole realm of biology. We have seen natural selection laid on its *Sterbebett*, and subsequently revived again in the most recent times to a remarkable degree of vigor. There can be no doubt that the old idea has great survival value.

The odd thing about the case, however, is that during all the years from 1859, when Darwin assembled in the *Origin of Species* a masterly array of concrete evidence for the reality of the struggle for existence and the process of natural selection, down to the present day, about all that biologists, by and large, have done regarding the idea is to talk and write. If ever an idea cried and begged for experimental testing and development, surely it was this one. Yet the whole array of experimental and statistical attempts in all these years to produce some significant new evidence about the nature and consequences of the struggle for existence is pitifully meager. Such contributions as those of Bumpus, Weldon, Pearson, and Harris are worthy of all praise, but there have been so very, very few of them. And there is surely something comic in the spectacle of laboratories overtly embarking upon the experimental study of evolution and carefully thereafter avoiding any direct and purposeful attack upon a pertinent problem, the fundamental importance of which Darwin surely established.

At the present time there is abundant evidence of an altered attitude; and particularly among the younger generation of biologists. The problem is being attacked, frontally, vigorously and intelligently. This renewed and effective activity seems to be due primarily to two things: first, the recrudescence of general interest in the problems of population, with the accompanying recognition that population problems are basically biological problems; and, second, the realization that the struggle for existence and natural selection are matters concerning the *dynamics of populations*, birth rates, death rates, interactions of mixed populations, etc. These things were recognized

v

and pointed out by Karl Pearson many years ago. His words, however, went largely unheeded for a long time. But in the last fifteen years we have seen more light thrown upon the problems of population by the work of such mathematicians as Lotka and Volterra, such statisticians as Yule, and such experimentalists as Allee and Park, than in the entire previous history of the subject. There can be no doubt of the fact that population problems now constitute a major focal point of biological interest and activity.

The author of the present treatise, Dr. G. F. Gause (who stands in the front rank of young Russian biologists, and is, it gives me great pleasure and satisfaction to say, a *protégé* of my old student and friend, Prof. W. W. Alpatov) makes in this book an important contribution to the literature of evolution. He marshals to the attack on the old problem of the consequences of the struggle for existence the ideas and the methods of the modern school of population students. He brings to the task the unusual and most useful equipment of a combination in his own person of thorough training and competence in both mathematics and experimental biology. He breaks new ground in this book. It will cause discussion, and some will disagree with its methods and conclusions, but no biologist who desires to know what the pioneers on the frontiers of knowledge are doing and thinking can afford not to read it. I hope and believe that it is but the beginning of a series of significant advances to be made by its brilliant young author.

RAYMOND PEARL.

Department of Biology,
School of Hygiene and Public Health,
The Johns Hopkins University.

AUTHOR'S PREFACE

This book is the outcome of a series of experimental investigations upon which I have been engaged for several years past. In these experiments an attempt was made to make use of all the advantages of the controlled study of the struggle for existence in the laboratory with various organisms low in the evolutionary scale. It became evident that the processes of competition between different species of protozoa and yeast cells are sometimes subject to perfectly definite quantitative laws. But it has also been found that these processes are extremely complicated and that their trend often do not harmonize with the predictions of the relatively simple mathematical theory. There is also a continued need for attack upon the problems of the struggle for existence along the lines of experimental physiology and biology, even though the results obtained cannot yet be adequately expressed in mathematical terminology.

I wish to express my sincere thanks to Professor W. W. Alpatov for interest in the experimental investigations and for valuable suggestions. To Professor Raymond Pearl I am deeply indebted for great assistance in the publication of this book, without which it could never have appeared before the American reader. I am also grateful to the Editors of *The Journal of Experimental Biology* and *Archiv für Protistenkunde* for permission to use material previously published in these periodicals.

G. F. GAUSE.

Laboratory of Ecology,
Zoological Institute,
University of Moscow,
Malaia Bronnaia 12, Kv. 33.
November, 1934

CONTENTS

CHAPTER I

CONTENTS

Chapter I

THE PROBLEM

(1) The struggle for existence is one of those questions which were very much discussed at the end of the last century, but scarcely any attempt was made to find out what it really represents. As a result our knowledge is limited to Darwin's brilliant exposition, and until quite recently there was nothing that we could add to his words. Darwin considered the struggle for existence in a wide sense, including the competition of organisms for a possession of common places in nature, as well as their destruction of one another. He showed that animals and plants, remote in the scale of nature, are bound together by a web of complex relations in the process of their struggle for existence. "Battle within battle must be continually recurring with varying success," wrote Darwin, and "probably in no one case could we precisely say why one species has been victorious over another in the great battle of life. . . . It is good thus to try in imagination to give to any one species an advantage over another. Probably in no single instance should we know what to do. This ought to convince us of our ignorance on the mutual relation of all organic beings; a conviction as necessary as it is difficult to acquire. All that we can do, is to keep steadily in mind that each organic being is striving to increase in a geometrical ratio; that each at some period of its life, during some season of the year, during each generation or at intervals, has to struggle for life and to suffer great destruction" ('59, pp. 56–57).

(2) But if our knowledge of the struggle for existence has since Darwin's era increased to an almost negligible extent, in other domains of biology a great progress has taken place in recent years. If we look at genetics, or general physiology, we find that a decisive advance has been made there, after the investigators had greatly simplified their problems and taken their stand upon the firm basis of experimental methods. The latter presents a particularly interesting example about which we would like to say a few words. We mean the investigations of the famous Russian physiologist J. P. Pavlov, who approached the study of the nervous activity of higher

1

animal by thoroughly objective physiological methods.   As Pavlov
('23) himself says, it is "the history of a physiologist's turning from
purely physiological questions to the domain of phenomena usually
termed psychical."   The higher nervous activity presents such a
complicated system, that without special experiments it is difficult
to obtain an objective idea of its properties.   It is known, firstly, that
there exist constant and unvarying reflexes or responses of the organism to the external world, which are considered as the especial "elementary tasks of the nervous system."   There exist besides other
reflexes variable to an extreme degree which Pavlov has named
"conditional reflexes."   With the aid of carefully arranged quantitative experiments in which the animal was isolated in a special chamber, all the complicating circumstances being removed, Pavlov
discovered the laws of the formation, preservation and extinction of
the conditional reflexes, which constitute the basis for an objective
conception of the higher nervous activity.   "I am deeply, irrevocably
and ineradicably convinced, says Pavlov, that here, on this way lies
the final triumph of the human mind over its problem—a knowledge
of the mechanism and of the laws of human nature."

(3) The history of the physiological sciences for the last fifty years
is very instructive, and it shows distinctly that in studying the
struggle for existence we must follow the same lines.   The complicated relationships between organisms which take place in nature
have as their foundation definite *elementary processes of the struggle for
existence.   Such an elementary process is that of one species devouring
another, or when there is a competition for a common place between a
small number of species in a limited microcosm.*   It is the object of the
present book to bring forward the evidence, firstly, that in studying
the relations between organisms in nature some investigators have
actually succeeded in observing such elementary processes of the
struggle for existence and, secondly, to present in detail the results of
the author's experiments in which the elementary processes have
been investigated in laboratory conditions.   The experiments made
it apparent that in the simplest case we can give a clear answer to
Darwin's question: why has one species been victorious over another
in the great battle of life?

(4) It would be incorrect to fall into an extreme and to consider
the complicated phenomena of the struggle for life in nature as simply
a sum of such elementary processes.   Leaving aside the existence in

nature of climatic factors which undergo rhythmical time-changes, the elementary processes of the struggle for life take place there amid a totality of most diverse living beings. This totality presents a *whole*, and the separate elementary processes taking place in it are still insufficient to explain all its properties. It is also probable that changes of the totality as a whole put an impress on those processes of the struggle for existence which are going on within it.

Nobody contests the complexity of the phenomena taking place in the conditions of nature, and we will not enter here into a discussion of this fact. Let us rather point out all the importance of studying the elementary processes of the struggle for life. At present our position is like that of biophysicists in the second half of last century. First of all it had been necessary to show that separate elementary phenomena of vision, hearing, etc., can be fruitfully studied by physical and chemical methods, and thereupon only did the question arise of studying the organism as a system constituting a whole.

(5) Certain authors at the close of last century occupied themselves with a purely logical and theoretical discussion of the struggle for existence. They proposed different schemata for classifying these phenomena, and we will now examine one of them in order to give just a general idea of those elementary processes of the struggle for life with which we will have to deal further on. To the first large group of these processes belongs the struggle going on between groups of organisms differing in structure and mode of life. In its turn this struggle can be divided into a direct and an indirect one. The struggle for existence is direct when the preservation of life of one species is connected with the destruction of another, for instance that of the fox and the hare, of the ichneumon fly and its host larva, of the tuberculosis bacillus and man. In the chapter devoted to the experimental analysis of the predator-prey relations we will turn our attention to this form of the struggle. In plants, as Plate ('13) points out, the direct form of the struggle for existence is found only in the case of one plant being a parasite of the other. Among plants it is the indirect competition, or the struggle for the means of livelihood that predominates; this has also a wide extension among animals. It takes place in the case when two forms inhabit the same place, need the same food, require the same light. We will later give a great deal of attention to the experimental study of indirect competition. To the second group of phenomena of the struggle for life belongs the

intraspecies struggle, between individuals of the same species, which in its turn can be divided into a direct and an indirect one.

(6) In this book we are interested in the struggle for existence among animals, and it is just in this domain that exact data are almost entirely lacking. In large compilative works one may meet an indication that the struggle for existence "owing to the absence of special investigations has become transformed into a kind of logical postulate," and in separate articles one can read that "our data are in contradiction with the dogma of the struggle for existence." In this respect zoologists are somewhat behind botanists, who have accumulated already some rather interesting facts concerning this problem.

What we know at present is so little that it is useless to examine the questions: what are the features common to the phenomena of competition in general, and what is the essential distinction between the competition of plants and that of animals, in connection with the mobility of the latter and the greater complexity of relations into which they enter? What interests us more immediately is the practical question: what are the methods by means of which botanists study the struggle for existence, and what alterations do these methods require in the domain of zoology?

First of all botanists have already recognized the necessity of having recourse to experiment in the investigation of competition phenomena, and we can quote the following words of Clements ('24, p. 5): "The opinions and hypotheses arising from observation are often interesting and suggestive, and may even have permanent value, but ecology can be built upon a lasting foundation solely by means of experiment. . . . In fact, the objectivity afforded by comprehensive and repeated experiment is the paramount reason for its constant and universal use."

However, the experiments so far made by botanists are devoted to the analysis of plant competition from the viewpoint of ontogenic development. The competition began when the young plantlets came in contact with one another, and all the decisive stages of the competition took place in the course of development of the same plants.

In such circumstances the question as to the causes of the victory of certain forms over others presents itself in the following aspect: By the aid of what morphological and physiological advantages of the process of individual development does one plant suppress another under the given conditions of environment? Clements has characterized this phenomenon in the following manner: "The beginning of competition is due to reaction when the plants are so spaced that the reaction of one affects the response of the other by limiting it. The initial advantage thus gained is increased by cumulation, since even a slight increase of the amount of energy or raw material is followed by corresponding growth and this by a further gain in response and reaction. A larger, deeper or more active root system enables one plant to secure a larger amount of the chresard, and the immediate reaction is to reduce the amount obtainable by the other. The stem and leaves of the former grow in size and number, and thus require more water, the roots respond by augmenting the absorbing surface to supply the demand, and automatically reduce the water content still further and with it the opportunity of a competitor. At the same time the correlated growth of stems and leaves is producing a reaction on light by absorption, leaving less energy available for the leaves of the competitor beneath it, while increasing the amount of food for the further growth of absorbing roots, taller stems and overshading leaves" (Clements, '29, p. 318).

(7) It is not difficult to see that for the study of the elementary processes of the struggle for existence in animals we need experiments of another type. We are interested in the processes of destruction and replacing of one species by another in the course of a *great number of generations*. We are consequently concerned here with the problem of an experimental study of the growth of mixed *populations*, depending on a very great number of manifold factors. In other words we have to analyze the properties of the *growing groups of individuals* as well as the interaction of these groups. Let us make for this purpose an artificial microcosm, i.e., let us fill a test tube with a nutritive medium and introduce into it several species of Protozoa consuming the same food, or devouring each other. If we then make numerous observations on the alteration in the number of individuals of these species during a number of generations, and analyze the factors that directly control these alterations, we shall be able to form an objective idea as to the course of the elementary processes of the struggle for existence. In short, *the struggle for existence among animals is a problem of the relationships between the components in mixed growing groups of individuals, and ought to be studied from the viewpoint of the movement of these groups.*

For the study of the elementary processes of the struggle for existence in animals we can have recourse to experiments of two types.
We can pour some nutritive medium into a test tube, introduce into
it two species of animals, and then neither add any food nor change
the medium. In these conditions there will be a growth of the number of individuals of the first and second species, and a competition
will arise between them for the common food. However, at a certain moment the food will have been consumed, or toxic waste products will have accumulated, and as a result the growth of the population will cease. In such an experiment a *competition* will take place
between two species *for the utilization of a certain limited amount of
energy*. The relation between the species we will have found at the
moment when growth has ceased, will enable us to establish in what
proportion this amount of energy has been distributed between the
populations of the competing species. It is also evident that one
can add to the species "prey" growing in conditions of a limited
amount of energy the species "predator," and trace the process of one
species being devoured by the other. Or, in the experiments of the
second type, *we need not fix the total amount of energy as a determined
quantity*, and only *maintain it at a certain constant level*, continually
changing the nutritive medium after fixed intervals of time. In such
an experiment we approach more closely to what takes place in the
conditions of nature, where the inflow of solar energy is maintained
at a fixed level, and we can study the process of competition for common food, or that of destruction of one species by another, in the
course of time intervals of any duration we may choose.

(8) Experimental researches will enable us to understand the
mechanism of the elementary process of the struggle for existence,
and we can proceed to the next step: to express these processes
mathematically. As a result we shall obtain coefficients of the
struggle for existence which can be exactly measured. The idea of a
mathematical approach to the phenomena of competition is not a
new one, and as far back as 1874 the botanist and philosopher Nägeli
attempted to give "a mathematical expression to the suppression of
one plant by another," taking for a starting point the annual increase
of the number of plants and the duration of their life. But this line
of investigation did not find any followers, and the experimental
researches on the competition of plants which have appeared lately

are as yet in the stage of nothing but a general analysis of the processes of ontogenesis.

In past years several eminent men were deeply conscious of the need for a mathematical theory of the struggle for existence and took definite steps in this domain. It often happened that one investigator was ignorant of the work of another but came to the same conclusions as his predecessor. Apparently every serious thought on the process of competition obliges one to consider it as a whole, and this leads inevitably to mathematics. A simple discussion or even a quantitative expression of data often do not suffice to obtain a clear idea of the relationships between the competing components in the process of their growth.

(9) About thirty years ago mathematical investigations of the struggle for existence would have been premature, or in any case subject to great difficulties, due to the absence of the needed preliminary data. Of late years, owing to the publication of a number of investigations, these difficulties have disappeared of themselves. What is it that these indispensable preliminary researches represent?

There is no doubt that a rational study of the struggle for existence among animals can be begun only after the questions of the multiplication of organisms have undergone a thoroughly exact quantitative analysis. We have mentioned that the struggle for existence is a problem of the relationships between species in mixed growing groups of individuals. We must therefore begin by analyzing the laws of growth of homogeneous groups consisting of individuals of one and the same species, and the competition between individuals in such homogeneous groups. During the second half of the last century and the beginning of the present much has been said about multiplication, and "equations of multiplication" have even been proposed of the following type: the coefficient of reproduction − the coefficient of destruction = number of adults. (Vermehrungsziffer − Vernichtungsziffer = Adultenziffer; see Plate ('13) p. 246.) Usually, however, things did not go any further, and no attempts were made to formulate exactly all these correlations. Recently the Russian geochemist, Prof. Vernadsky, has thus characterized from a very wide viewpoint the phenomena of multiplication of organisms ('26, p. 37 and foll.): "The phenomena of multiplication attracted but little the attention of biologists. But in it, partly unnoticed by the naturalists themselves, several empirical generalizations became established to which we have become so accustomed that they appear to us almost self-evident.

"Among these generalizations the following must be recorded. Firstly, *the multiplication of all organisms can be expressed by geometric progressions.* This can be evaluated by a uniform formula:

$$2^{bt} = N_t$$

where t is time, b the exponent of progression and N_t the number of individuals existing owing to multiplication at a certain time t. Parameter b is characteristic for every kind of living being. In this formula there are included no limits, no restrictions either for t, for b, or for N_t. The process is conceived as infinite as the progression is infinite.

"This infinity of the possible multiplication of organisms can be considered as the subordination of *the increase of living matter in the biosphere to the rule of inertia.* It can be regarded as empirically established that the process of multiplication is retarded in its manifestation only by external forces; it dies off with a low temperature, ceases or becomes weaker with an insufficiency of food or respiration, with a lack of room for the organisms that are being newly created. In 1858 Darwin and Wallace expressed this idea in a form that had been long clear to naturalists who had gone into these phenomena, for instance, Linnaeus, Buffon, Humboldt, Ehrenberg and von Baer: if there are no external checks, every organism can, but at a different time, cover the entire globe by its multiplication, produce a progeny equal in size to the mass of the ocean or of the earth's crust.

"The rate of multiplication is different for every kind of organisms in close connection with their size. *Small organisms, that is organisms weighing less, at the same time multiply much more rapidly than large organisms (i.e., organisms of a great weight).*

"In these three empirical generalizations the phenomena of multiplication are expressed without any consideration of time and space or, more precisely, in geometrical homogeneous time and space. In reality life is inseparable from the biosphere, and we must take into consideration terrestrial time and space. Upon the earth organisms live in a limited space equal in dimensions for them all. They live in a space of definite structure, in a gaseous environment or a liquid environment penetrated by gases. And although to us time appears unlimited, the time taken up by any process which takes place in a limited space, like the process of multiplication of organisms, cannot be unlimited. It also will have a limit, different for every kind of organisms in accordance with the character of its multiplication. The inevitable consequence of this situation is a limitation of all the parameters which determine the phenomena of multiplication of organisms in the biosphere.

"For every species or race there is a maximal number of individuals which can never be surpassed. This maximal number is reached when the given species occupies entirely the earth's surface, with a maximal density of its occupation. This number which I will henceforth call the 'stationary number of the homogeneous living matter' is of great significance for the evaluation of the geochemical influence of life. The multiplication of organisms in a given volume or on a given surface must proceed more and more slowly, as the number of the individuals already created approaches the stationary number."

These general notions on the multiplication of organisms have lately received a rational quantitative expression in the form of the logistic curve discovered by Raymond Pearl and Reed in 1920. The logistic law mathematically expresses the idea that in the conditions of a limited microcosm the potentially possible "geometric increase" of a given group of individuals at every moment of time is realized only up to a certain degree, depending on the unutilized opportunity for growth at this moment. As the number of individuals increases, the unutilized opportunity for the further growth decreases, until finally the greatest possible or saturating population in the given conditions is reached. The logistic law has been proved true as regards populations of different animals experimentally studied in laboratory conditions. We shall have an opportunity to consider all these problems more in detail further on. Let us now only note that the rational quantitative expression of growth of groups consisting of individuals of the same species represents a firm foundation for a further fruitful study of competition between species in mixed populations.

(10) Apart from a great progress as regards the mathematical expression of the multiplication of organisms, an important advance has taken place in the theory of competition itself. The first step in this direction was made in 1911 by Ronald Ross, who at this time was interested in the propagation of malaria. Considering the

process of propagation Ross came to the conclusion that he was dealing with a peculiar case of a struggle for existence between the malaria plasmodium and man with a participation of the mosquito. Ross formulated mathematically an equation of the struggle for existence for this case, which closely approached in its conception those equations of the struggle for existence which the Italian mathematician Volterra proposed in 1926 without knowing the investigations of Ross.

Whilst Ross was working on the propagation of malaria the American mathematician Lotka ('10, '20a) examined theoretically the course of certain chemical reactions, and had to deal here with equations of the same type. Later on Lotka became interested in the problem of the struggle for existence, and in 1920 he formulated an equation for the interaction between hosts and parasites ('20b), and gave a great deal of interesting material in his valuable book, *Elements of Physical Biology* ('25). Without being acquainted with these researches the Italian mathematician Vito Volterra proposed in 1926 somewhat similar equations of the struggle for existence. At the same time he advanced the entire problem considerably, investigating for the first time many important questions of the theory of competition from the theoretical point of view. Thus three distinguished investigators came to the very same theoretical equations almost at the same time but by entirely different ways. It is also interesting that the struggle for existence only began to be experimentally studied after the ground had been prepared by purely theoretical researches. The same has already happened many times in the fields both of physics and of physical chemistry: let us recollect the mechanical equivalent of heat or Gibbs' investigations.

(11) The study of the struggle for existence will undoubtedly rapidly progress in the future, but it will have to overcome a certain gap between the investigations of contemporary biologists and mathematicians. There is no doubt that the struggle for existence is a biological problem, and that it ought to be solved by experimentation and not at the desk of a mathematician. But in order to penetrate deeper into the nature of these phenomena we must combine the experimental method with the mathematical theory, a possibility which has been created by the brilliant researches of Lotka and Volterra. This combination of the experimental method with the quantitative theory is in general one of the most powerful tools in the hands of contemporary science.

The gap between the biologists and the mathematicians represents a significant obstacle to the application of the combined methods of research. Mathematical investigations independent of experiments are of but small importance due to the complexity of biological systems, narrowing the possibilities of theoretical work here as compared with what can be admitted in physics and chemistry. We are in complete accord with the following words of Allee ('34): "Mathematical treatment of population problems is necessary and helpful, particularly in that it permits the logical arrangement of facts and abbreviates their expression by the use of a sort of universal shorthand, but the arrangement and statement may lead to error, since for the sake of brevity and to avoid cumbersome expressions, variables are omitted and assumptions made in the mathematical analyses which are not justified by the biological data. Certainly there is room for the mathematical attack on population problems, but there is also continued need for attack along the lines of experimental physiology, even though the results obtained cannot yet be adequately expressed in mathematical terminology."

CHAPTER II

THE STRUGGLE FOR EXISTENCE IN NATURAL
CONDITIONS

(1) Before beginning any experimental investigation of the elementary processes of the struggle for existence we must examine what is the state of our knowledge of the phenomena of competition in nature. The regularities which it has been possible to ascertain there, and the ideas which have been expressed in their discussion, will help us to formulate correctly certain fundamental requirements for further experimental work.

In thorough field observations the fact which strikes the investigator most of all is the extreme complexity of the communities of organisms, and at the same time their possession of a definite structure. On the one hand they undergo changes under the influence of external environment, and on the other the slightest changes of some components produce an alteration of others and lead to a whole chain of consequences. It is difficult here to arrive at a sufficiently clear understanding of the processes of the struggle for existence. Elton writes for instance: "We do not get any clear conception of the exact way in which one species replaces another. Does it drive the other one out by competition? and if so, what precisely do we mean by competition? Or do changing conditions destroy or drive out the first arrival, making thereby an empty niche for another animal which quietly replaces it without ever becoming 'red in tooth and claw' at all? Succession brings the ecologist face to face with the whole problem of competition among animals, a problem which does not puzzle most people because they seldom if ever think out its implications at all carefully. At the present time it is well known that the American grey squirrel is replacing the native red squirrel in various parts of England, but it is entirely unknown why this is occurring, and no good explanation seems to exist. In ecological succession among animals there are thousands of similar cases cropping up, practically all of which are as little accounted for as that of the squirrels" ('27, p. 27–28). All this suggests that an analysis must be made of comparatively simple desert or Arctic communities where the number of

12

components is small. Such a tendency to examine certain *elementary phenomena* is clearly seen in the following words of a Russian zoologist, N. Severtzov, written as far back as 1855: "It seems to me that the study of animal groupings in small areas, the study of these *elementary faunas* is the firmest point of support for drawing conclusions about the general laws regulating the distribution of animals on the globe."

However, besides this first possibility of studying competition phenomena among a small number of components, an active intervention into natural conditions by means of biotic experiments may also be very important. Among such experiments the most frequent ones consist in the transportation of animals into countries new to them, which commonly leads to a great number of highly interesting proc-

TABLE I

Number of fir trunks on a unit of surface under different conditions

From Sukatschev ('28)

TYPE OF LIFE CONDITIONS	20 YEARS AGE		60 YEARS AGE	
	Predominant trunks	Oppressed trunks	Predominant trunks	Oppressed trunks
I	5600	—	1300	640
II	5850	—	1600	680
III	6620	—	1950	650
IV	7480	—	2280	720
V	8400	—	2780	760

esses of the struggle for existence (Thomson, '22). The second type of biotic experiments is an "exclusion" of the animal from a certain community. Further on we give some examples of the struggle for existence observed by such methods, but so far none of them have been sufficiently studied.

(2) It fell to the lot of botanists to have to deal with the simplest conditions of competition, and they arrived at a very instructive conception of the *intensity of the struggle for existence*. Foresters were the first to be confronted with the question of competition when they began to estimate the diminution in the number of tree trunks accompanying forest growth in different conditions of environment. They characterize the struggle for existence by the percentage decrease in the number of individuals on a unit of surface in a certain unit of

time.   At first sight one might think that the better the conditions of existence the less active is the struggle for life, and the greater the number of trunks that can survive with age on a unit of surface.   Let us, however, look at the data of the foresters.   For an example we will give in Table I the number of the fir trunks in the government of Leningrad (Northern Russia) corresponding to five different types of life conditions (Type I represents the best soil and ground conditions; V, the worst ones).

These data show, contrary to our expectations, that the better are the soil and ground conditions, the more active is the struggle for life, or in other words the smaller the number of trunks remaining on a unit of surface and, consequently, the greater the percentage of those which perish.   If we think out this phenomenon, it becomes quite understandable: the more favorable the environment is for the plants' existence, the more luxuriant will be the development of each plant, the sooner will the tops of the trees begin to close above, and the earlier the oppressed individuals become isolated.   Also, in better conditions of existence, every individual in the adult state will be more developed and occupy a greater space, but the individuals will be fewer in number.   Investigations show that this is a general rule for all the forest species (Sukatschev, '28, p. 12).

Similar data were obtained by Sukatschev ('28) in experiments with the chamomile, *Matricaria inodora*, on fertilized and non-ferti- lized soil.   In counting up the individuals remaining at the end of summer (August 17), the following decrease of the original number of individuals was ascertained (see Table II and Fig. 1).

Here likewise in better conditions of existence competition pro- ceeds with greater intensity, and the per cent of individuals which perish is greater.

The results obtained by botanists are certainly characteristic for the ontogenetic development of plants, but at the same time they give us an approach to the quantitative appreciation of the intensity of the struggle for existence, the whole significance of which was already clearly understood by Darwin.   In the next chapter we shall consider the struggle for life in animals, and there, using entirely different methods, we shall endeavor to formulate quantitatively the intensity of this struggle.

(3) In field observations the question often arises as to the struggle for existence in mixed populations, about which Darwin wrote: "As

the species of the same genus usually have, though by no means invariably, much similarity in habits and constitution, and always in structure, the struggle will generally be more severe between them, if they come into competition with each other, than between the species of distinct genera." Lately, botanists have tried to approach this problem experimentally. It became evident that, actually, in a number of cases competition is keenest when the individuals are most

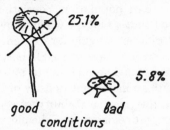

Percentage of perished individuals

25.1%

5.8%

good   Bad
      conditions

FIG. 1. Intensity of the struggle for existence in the chamomile, *Matricaria inodora*, on fertilized and non-fertilized soil (dense culture).

TABLE II

Decrease of the number of individuals in the chamomile (Matricaria inodora) expressed in percentage of the initial number

From Sukatschev ('28)

	PERCENTAGE
Dense culture (3 x 3 cm.):	
Non-fertilized soil	5.8
Fertilized soil	25.1
Culture of middle density (10 x 10 cm.):	
Non-fertilized soil	0.0
Fertilized soil	3.1

similar. The more unlike plants are, the greater difference in their needs, and hence some adjust themselves to the reactions of others with little or no disadvantage. This similarity must rest upon vegetation or habitat form, and not merely upon systematic position (Clements, '29). Researches on competition in mixed populations consisting of different kinds of cultivated plants were undertaken by many investigators (e.g., Montgomery, '12). Particularly interesting data concerning wild-growing plants have been recently published by Sukatschev ('27) in his "Experimental studies on the struggle for existence between biotypes of the same species." First of all he studied the competition between local biotypes of the plant, *Taraxacum officinale* Web., from the environs of Leningrad. These biotypes were cultivated in similar conditions with a fixed distance between the individuals, and the experiments led Sukatschev to the following conclusions: (a) One must rigorously distinguish the conditions of the struggle for existence in a pure population, formed by a single biotype, and in a mixed population, consisting of various biotypes. (b) It is to be noted that a biotype which shows itself to be the most resistant in an *intra*biotic struggle for existence, may turn out to be the weakest one in an *inter*biotic struggle between different biotypes of the same species. (c) The increase in mutual influence of plants upon each other with an increase in density of the plant cultures, may completely reverse the relative stability of separate biotypes in the process of the struggle for existence. The biotypes yielding the greatest percentage of survivors under a small density of cultivation may occupy the last place in this respect in conditions of a dense culture. This can be illustrated by Table III.

If we arrange the biotypes mentioned in Table III according to decreasing stability, we shall find that in the conditions of a *not dense, pure culture*: $C > A > B$, i.e., the biotype C gives the smallest percentage of non-survivals and is the *most resistant*, whilst the biotype B is the weakest of all. In *dense pure cultures* the relations are entirely different: $B > A > C$, i.e., the biotype B is the most stable one. Lastly, for *dense*, but *mixed cultures*, we have: $C > A > B$. Almost similar data have been obtained by Montgomery ('12) in studying the competition between two races of wheat.

In another series of experiments Sukatschev ('27) studied the struggle for existence between biotypes of various geographical origin (from various parts of U. S. S. R.), and inferred the following: (a) Judging by the percentage of non-survivals, one can say that in pure, as well as in mixed not-dense cultures, the dying-off is chiefly due to the influence of physico-geographical factors. Therefore, the biotypes originating from geographical regions strongly differing from a given region in their climate, turn out to be less resistant, as compared with the local biotypes. But these relations can change under the influence of aggregation. (b) In dense mixed cultures, if we are to judge by the percentage of perished individuals, it is not the local biotypes that appear the most resistant in the struggle for existence, but those introduced from other regions.   (c) The struggle for existence in mixed cultures of various biotypes is not so keen as that in pure cultures of separate biotypes with the same density.

The analysis of the struggle for existence in mixed populations of plants is now only at its very beginning.   The exact data are few in number, but there exist numerous observations on the stratified distribution of plants (see Alechin, '26), considering these strata as a result of complex processes of competition and adaptation of the plants to one another in mixed cultures.

(4) Plants are also very favorable for the study of the influence of

TABLE III

Percentage of eliminated individuals in three biotypes, A, B and C of Taraxacum officinale

From Sukatschev ('27)

	RACE	SPARCE CULTURES	DENSE CULTURES
Pure cultures....................... {	A	22.9	73.2
	B	31.1	51.1
	C	10.3	75.9
Mixed cultures..................... {	A	16.5	77.4
	B	22.1	80.4
	C	5.5	42.0

environment upon the struggle for existence in mixed populations. DeCandolle was already interested in this question in 1820, but as regards exact experimental researches very little has been done until quite recently, when the works of Tansley ('17) and others appeared. These investigations show that in pure cultures two species can grow for a certain time on various soils, but each of the species has an advantage over the others in particular soil conditions.   Therefore in mixed population in some soils the first species displaces the second, but in others the second species displaces the first.[1]   The actual rela-

---

[1] Among animals such observations have been recently made by Timofeeff-Ressovsky ('33), who studied the competition between the larvae of *Drosophila melanogaster* and *Drosophila funebris* under different temperatures.   We can mention also certain interesting data of Beauchamp and Ullyott ('32): "When

tions existing here are, however, somewhat complicated (see Braun-Blanquet, '28).

The problem of the influence of environment on competition presents considerable interest, but as yet what we know is very meager. In the majority of cases it is observations of a qualitative character, of which we can give an example here: "The root-systems of the vegetation in the steppes of Southern Russia form, according to Patchossky, three strata. The uppermost one consists of short roots belonging to annual plants which vegetate for a short time. The second, deeper-lying stratum belongs to the essential plants of the steppe vegetable covering, the Gramineae. The third, deepest stratum consists of the vertical stem-like roots of perennial dicotyledons (among them the steppe *Euphorbia*). Usually, the second gramineous stratum dominates. When, however, an immoderate pasturing takes place in a given locality, the gramineous covering begins to suffer and does not produce a vigorous root-system. Atmospheric precipitation can now penetrate to those soil horizons where roots of the dicotyledons are situated, and the latter begin to dominate. As a result appears an unbroken vegetable covering consisting of *Euphorbia*. Analogous results take place in case of increase in yearly atmospheric precipitation. In this case, although the water is energetically absorbed by the second gramineous root stratum, the rainfall is so considerable that a great part of the water penetrates deeper, contributing to the development of dicotyledonous plants. The large dicotyledons act depressingly upon the Gramineae, and they change places in respect to their domination" (Alechin, '26).

(5) The part which the quantitative relations between species at the beginning of their struggle play in the outcome of competition presents an interesting problem. Botanists do not possess exact quantitative data bearing on this question, and one meets only with considerations of the following kind: When new soils are colonized, if the species concerned do not sharply differ in their capacity for spreading, it mostly depends on chance which species colonizes the given area first. But this chance determines the further colonizing

---

Planaria montenegrina and *Pl. gonocephala* occur in competition with each other, temperature is the factor which governs the relative success and efficiency of the two species. *Pl. montenegrina* is the more successful at temperatures below 13–14°C. Above these temperatures *Pl. gonocephala* is the more efficient form."

of the given locality. Even when the species that has first established itself is somewhat weaker than another species in the same habitat, it can for a comparatively long time resist its stronger competitor simply because it was the first to occupy this place. Only in case of a considerable weakness of the first comer will its domination be merely a temporary one, and the effect of the first accidental appearance will be rapidly eliminated (E. Warming ('95), Du-Rietz ('30)).

(6) Let us recapitulate briefly our discussion up to this point. Botanists have endeavored to investigate the struggle for existence by experimentation and under simplified conditions, but they are only beginning to analyze these phenomena. Their experiments are commonly limited to the process of ontogenetic development, and in only a few cases, chiefly concerned with competition in cereals, has displacement of some forms by others been traced through a series of generations (Montgomery ('12) and others). As concerns animals we have simply no exact data, and can only mention a few general principles which have been developed by zoologists in connection with the phenomena of competition.

One of these ideas is that of the "niche" (see Elton, '27, p. 63). A niche indicates what place the given species occupies in a community, i.e., what are its habits, food and mode of life. It is admitted that as a result of competition two similar species scarcely ever occupy similar niches, but displace each other in such a manner that each takes possession of certain peculiar kinds of food and modes of life in which it has an advantage over its competitor. Curious examples of the existence of different niches in nearly related species have recently been obtained by A. N. Formosov ('34). He investigated the ecology of nearly related species of terns, living together in a definite region, and it appeared that their interests do not clash at all, as each species hunts in perfectly determined conditions differing from those of another. This once more confirms the thought mentioned earlier, that the intensity of competition is determined not by the systematic likeness, but by the similarity of the demands of the competitors upon the environment. Further on we shall endeavor to express all these relations in a quantitative form.

(7) The above mentioned observations of A. N. Formosov on different niches in nearly related species of terns can be given here with more detail, as the author has kindly put at our disposal the following materials from his unpublished manuscript: According to

the observations in 1923, the island Jorilgatch (Black Sea) is inhabited by a nesting colony of terns, consisting of many hundreds of
individuals.   The nests of the terns are situated close to one another,
and the colony presents a whole system.   The entire mass of individuals in the colony belongs to four species (sandwich-tern, *Sterna
cantiaca*; common-tern, *S. fluviatilis*; blackbeak-tern, *S. anglica*; and
little-tern, *S. minuta*), and together they chase away predators (hen-
harriers, etc.) from the colony.   However, as regards the procuring
of food, there is a sharp difference between them, for every species
pursues a definite kind of animal in perfectly definite conditions.
Thus the sandwich-tern flies out into the open sea to hunt certain
species of fish.   The blackbeak-tern feeds exclusively on land, and
it can be met in the steppe at a great distance from the sea-shore,
where it destroys locusts and lizards.   The common-tern and the
little-tern catch fish not far from the shore, sighting them while flying
and then falling upon the water and plunging to a small depth.   The
light little-tern seizes the fish in shallow swampy places, whereas the
common-tern hunts somewhat further from the shore.   In this manner these four similar species of tern living side by side upon a single
small island differ sharply in all their modes of feeding and procuring
food.

(8) Another ecological notion is also important in connection with
our experiments.   We have in view the degree of isolation of the
microcosm.   The point is that our experimental researches have been
mainly made in isolated microcosms, i.e., in test tubes filled with
nutritive medium and stopped with cotton-wool.   It must be re-
membered that the degree of isolation of different communities in
natural conditions is very different.   Such a system as a lake is almost
isolated, but at times some of the animals inhabiting it go on land.
An oasis in the desert would also seem to be isolated, but for instance
some of the species of birds fly away for the winter, and consequently
there is no real isolation.   The habitats not so sharply separated from
the surrounding life-area are, therefore, still less isolated.   All this
emphasizes the idea already expressed, that the regularities observed
in isolated microcosms hold true only under certain fixed conditions,
and are not sufficient to explain all the complicated phenomena taking place in nature.   We shall have an opportunity to appreciate the
rôle of this factor when experimentally studying the predator-prey
relations.

Let us note another important circumstance connected with competition. This phenomenon can be particularly pronounced during a periodical food shortage connected with certain seasons, etc., whilst at another time with an abundance of food it will scarcely take place. This fact has frequently been pointed out in various discussions of the struggle for existence.

(9) It remains but to give some examples of the struggle for existence among animals in order to show with what problems the zoologists have to deal, and how difficult it is to apply here exact quantitative methods. An instructive example of competition among fishes has been recently described by Kashkarov ('28). It concerns the supplanting of *Schizothorax intermedius* by wild carp, *Cyprinus albus* L., in lakes of Middle Asia. Wild carp were introduced into the lake Bijly Kul in 1909. Before that only *Schizothorax intermedius* with white-fish (*Leuciscus* sp.) inhabited this lake. Formerly *Schizothorax intermedius* were very numerous, but after the introduction of wild carp their quantity diminished considerably. As an indicator of the relatively small number of *Schizothorax intermedius* the following data on the catch may serve: on May 15, 1926, 19 carp and 1 *Schizothorax* were caught in two nets; on May 16, 1926, in the same place the catch was 24 carp and 2 *Schizothorax*. *Schizothorax* keeps chiefly to the south-western part of the lake, where there are stones making the casting of nets difficult. Now *Schizothorax* is disappearing even there, as wild carp devour its spawn. The quantity of white-fish also decreases because carp devour its young. The particular interest of this example lies in the fact that a new species not found in a given microcosm before (*Cyprinus albus* L.) was introduced, and in this way a direct proof of one species displacing another was obtained.

(10) Processes of this kind can often be observed when fish are introduced into waters to which they are new. Professor G. C. Embody writes in a letter recently received: "Concerning the competition between different species of fishes we have two cases in particular in the eastern United States. The European carp was introduced in the '70's, and has now in many streams and lakes multiplied to such an extent that several native species are found in greatly diminished number. This has probably been due to the high reproductive capacity of the carp, food competition, destruction of weed beds by carp, and the fact that very few of them are captured.

Carp are not used as extensively for food in America as in Europe and in our smaller lakes are not generally fished for commercially. The other case is the introduction of the perch (*Perca flavescens*) into certain lakes in the Adirondacks and in Maine, which were naturally populated with the trout (*Salvelinus fontinalis*). The competition for food is believed to be one of the causes for the decrease in the number of the trout."

"These cases are both matters of general observation. I do not know of any papers describing them nor in fact, dealing with this subject in American waters."

(11) Another example of competition is the replacing of one species of cray-fish by another in certain waters of Middle Russia. Some observations on this were made by Kessler ('75) and recently by Birstein and Vinogradov ('34). Two species of cray-fish inhabit the waters of European Russia: the broad-legged (*Potamobius astacus* L.) and long-legged (*Potamobius leptodactylus* Esch.). The broad-legged cray-fish is distributed in the western part, and the long-legged in the south-eastern one, but the areas of their distribution largely overlap one another. It is observed that the long-legged cray-fish displaces the broad-legged one and spreads gradually more and more to the west. It has been possible to establish this replacement with particular distinctness in White Russia (in the western part of U. S. S. R.). The cray-fish are found there in lakes isolated from each other, and most of the lakes are inhabited only by the broad-legged cray-fish. In some cases long-legged cray-fish were put into such lakes from other waters. As a result the broad-legged cray-fish began to decrease, and *finally disappeared completely* leaving the lake populated exclusively by the long-legged species. The following examples can be given. (I) Black lake (White Russia) was populated only by the broad-legged cray-fish. In 1906, 500 specimens of long-legged cray-fish were introduced, and now (1930) only this species remains. (II) Forest lake (same region). Up to 1920 there were no long-legged cray-fish there. Later on they were introduced, and at present (1930) there is a considerable number of this species. The causes why one species of cray-fish is replaced by another have scarcely been studied.

(12) Curious are the observations reported by Goldman ('30) on the competition among predators belonging to different species. Thus, according to a resident of Telegraph Creek near the Stikine River, Canada, no coyotes were known in that section prior to 1899.

About that time, however, they came in, apparently following the old goldrush trail, probably attracted by the hundreds of dead horses along it. The invasion of Alaska seems destined to continue until coyotes have extended their range over practically all of the territory.

It has been found in Alaska that the coyotes kill many foxes. Since the coyotes have increased the foxes have decreased alarmingly. In some sections practically none are believed to be left. In many cases a family or entire colony of foxes are run out of their dens or are both run out and killed by coyotes which then use the dens themselves. Wolves are well known to have committed similar depredations, but their killing is not so extensive as is that of the coyotes. Serious as are the depredations of wolves throughout most of Alaska, the damage done to game and fur bearing animals by coyotes is not only far greater but is rapidly increasing in extent. How far the coyotes will hold back the normal development in the periodical increase of the snowshoe rabbits and the ptarmigan which are important items in the food supply of fur bearers, especially the lynx and fox, it is impossible even to approximate.

(13) The examples just mentioned show that the introduction of aquatic animals into waters to which they are new, or the penetration of land animals into new regions, often lead to very interesting processes of competition. We would now like to say a few words about the *direct* struggle for existence, in which one species devours another. In this case it is very important to ascertain the exact numerical relation between the population of the devoured species and that of the devouring one, and this can often be attained by changing their relative quantities. One of the outstanding facts is the increase of the deer accompanying the destruction of wolves, foxes, etc., by early settlers in Illinois which, according to Wood, has been recently reported by Shelford ('31). Wood depicts a continuous decrease in wolves and wildcats from the beginning of settlement to their practical extinction. When the wolf population was reduced to about one-half, the deer increased rapidly for a little less than 10 years, reaching a large maximum of about three times the original number. Unfortunately we have no *exact* data on the change of the numerical relation between the wolves and the deer, but such processes are in any case of great interest.

The change of the numerical relations between the predator and the prey sometimes takes place as a consequence of a mass appearance of the prey in years especially favorable for its multiplication. This frequently happens with wild mice, and it gives us the possibility of tracing the process of their being devoured by the predator. In this connection we may mention the following observations recently made by Kalabuchov and Raewski ('33) in the North Caucasus: "The picture of the destruction of mice by different predators is a curious one. At the beginning of the destruction about the same number of rodents is devoured daily. But as the density of rodents diminishes it becomes more and more difficult to catch them, and the number of mice devoured gradually decreases. Finally a time comes when the relation between the density of the rodents, the presence of cover or refuge (burrows, vegetation, etc.) and the biological

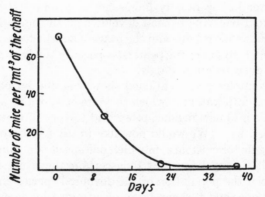

FIG. 2. The destruction of mice by different predators in heaps of chaff

peculiarities of the predators becomes such that the latter can devour the rodents only in rare cases. In this way the number of the rodents remains about constant.

"The data on the change in the number of mice near the village of Kambulat are a good illustration of this regularity. Having established that the destruction of mice in this locality was due to owls, polecats and other predators that devoured them, we obtained the following picture of the change in the number of the rodents in heaps of chaff (Fig. 2). This figure shows that with a density of 2.5 − 0.8 mice per mt^3 of chaff we have conditions in which the destruction of the rodents by the predators became so rare that their number scarcely varied."

Certain interesting observations have been also recently made by ecological entomologists (Payne, '33, '34) on the host-parasite balance. It is possible to trace the process of the destruction of population of the moth *Ephestia* by the hymenopterous parasite *Microbracon* in the laboratory, and it appears from the observations that a great many factors are important for the process of their interaction, and particularly various relations between "susceptible" stage of host and "effective" stage of parasite. The beginning of the theoretical investigation of this case has been given by the interesting papers of Bailey ('31, '33) and Nikolson ('33). For such investigations, however, the populations of unicellular organisms are somewhat more convenient.

(14) The few examples given above show sufficiently that the processes of the struggle for existence among animals are of extreme importance from a practical point of view. They are sharply outlined in isolated microcosms and therefore there is nothing surprising if they have attracted the particular attention of the workers in the domain of fishery. Lately the problem of the relations between predatory and non-predatory fish has been discussed by Italian authors (D'Ancona ('26, '27), Marchi ('28, '29), Brunelli ('29)). D'Ancona collected the data of a statistical inspection of the fish-markets in Triest, Venice and Fiume for several years. He claims that the diminished intensity of fishing during the war-period (1915–1920) has caused a comparative increase of the number of predatory fish. He therefore reasons that fishing of normal intensity causes a relative diminution of the number of predatory fish and a comparative increase of the non-predatory ones. But his data are not convincing and indeed Bodenheimer ('32) has recently shown that such variations in the fish population existed before and after the war. They are apparently not connected with the intensity of fishing but probably are the results of certain changes of the environment. However it may be, the material collected by D'Ancona stimulated the highly interesting mathematical researches on the struggle for existence of Vito Volterra. Although his mathematical theories are not confirmed in any way by D'Ancona's statistical data, the importance of Volterra's methods as a new and powerful tool in the analysis of biological populations admits of no doubt.

(15) In concluding this descriptive chapter of our book let us note the following picture of the struggle for existence in nature. It is only in the domain of botany that these processes are coming to be

investigated from a certain general viewpoint as (1) intensity of competition, (2) competition in mixed populations, (3) the influence of environment upon competition in mixed cultures, and (4) the rôle of the quantitative relations between species at the beginning of their struggle.   Among animals the processes of the struggle for existence are much more complex, and as yet one cannot speak of any general principles.   In this connection an investigation of the elementary processes of the struggle for life in strictly controlled laboratory conditions is here particularly desirable, and the material just presented will be of great help to us in the choice and arrangement of the corresponding experiments.

Chapter III

THE STRUGGLE FOR EXISTENCE FROM THE POINT OF VIEW OF THE MATHEMATICIANS

I

(1) In this chapter we shall make the acquaintance of the astonishing theories of the struggle for existence developed by mathematicians at a time when biologists were still far from any investigation of these phenomena and had but just begun to make observations in the field.

The first attempt at a quantitative study of the struggle for existence was made by Sir Ronald Ross ('08, '11). He undertook a theoretical investigation of the propagation of malaria, and came to conclusions which are of great interest for quantitative epidemiology and at the same time constitute an important advance in the understanding of the struggle for existence in general. Let us examine the fundamental idea of Ross. Our object will be to give an analysis of the propagation of malaria in a certain locality under somewhat simplified conditions. We assume that both emigration and immigration are negligible, and that in the time interval we are studying there is no increase of population or in other words the birth rate is compensated by the death rate. In such a locality a healthy person can be infected with malaria, according to Ross, if all the following conditions are realized: (1) That a person whose blood contains a sufficient number of gametocytes (sexual forms) is living in or near the locality. (2) That an Anopheline capable of carrying the parasites sucks enough of that person's blood. (3) That this Anopheline lives for a week or more afterwards under suitable conditions—long enough to allow the parasites to mature within it, and (4) that it next succeeds in biting another person who is not immune to the disease or is not protected by quinine. The propagation of malaria in such a locality is determined in its general features by two continuous and simultaneous processes: on the one hand the number of new infections among people depends on the number and infectivity of the mosquitoes, and at the same time the infectivity of the mosquitoes is connected with the number of people in the given locality and the frequency of the sickness among them. Ross has expressed in mathematical terms this uninterrupted and simultaneous dependence of the infection of the first component on the second, and that of the second on the first, with the aid of what is called in mathematics simultaneous differential equations. These equations are very simple and we shall examine them at once. In a quite general form they can be represented as follows:

$$\begin{Bmatrix} \text{Rate of increase of} \\ \text{affected individuals} \\ \text{among the human} \\ \text{population} \end{Bmatrix} = \begin{Bmatrix} \text{New infections} \\ \text{per unit of time.} \\ \text{Depends on in-} \\ \text{fected mosqui-} \\ \text{toes} \end{Bmatrix} - \begin{Bmatrix} \text{Recoveries} \\ \text{per unit of} \\ \text{time} \end{Bmatrix} *$$

$$\begin{Bmatrix} \text{Rate of increase of} \\ \text{infected individuals} \\ \text{among the mosquito} \\ \text{population} \end{Bmatrix} = \begin{Bmatrix} \text{New infections} \\ \text{per unit of time.} \\ \text{Depends on the} \\ \text{infected humans} \end{Bmatrix} - \begin{Bmatrix} \text{Deaths of} \\ \text{infected} \\ \text{mosquitoes} \\ \text{per unit of} \\ \text{time} \end{Bmatrix} † \quad \dots (1)$$

(2) Translating these relations into mathematical language we shall obtain the simultaneous differential equations of Ross. Let us introduce the following notation (that of Lotka (51)):

$p =$ total number of human individuals in a given locality.
$p^1 =$ total number of mosquitoes in a given locality.
$z =$ total number of people infected with malaria.
$z^1 =$ total number of mosquitoes containing malaria parasites.
$fz =$ total number of infective malarians (number of persons with gametocytes in the blood; a certain fraction of the total number of malarians).
$f^1z^1 =$ total number of infective mosquitoes (with matured parasites; a certain fraction of the total number of mosquitoes containing parasites).
$r =$ recovery rate, i.e., fraction of infected population that reverts to noninfected (healthy) state per unit of time.

---

* The human death rate is not taken into consideration. To simplify the situation we assume that it is negligible and counterbalanced by the birth rate.

† We assume that the recovery of the infected mosquitoes does not occur.

M^1 = mosquito mortality, i.e., death rate per head per unit of time.

t = time.

If a single mosquito bites a human being on an average b^1 times per unit of time, then the f^1z^1 infective mosquitoes will place $b^1f^1z^1$ infective bites on human beings per unit of time. If the number of people not infected with malaria is $(p - z)$, than taken in a relative form as $\dfrac{p - z}{p}$ it will show the relative number of healthy people in the total number of individuals of a given locality. Therefore, out of a total number of infective bites, equal per unit of time to $b^1f^1z^1$, a definite fraction equal to $\dfrac{p - z}{p}$ falls to the lot of healthy people, and the number of infective bites of healthy people per unit of time will be equal to:

$$b^1 f^1 z^1 \frac{p - z}{p} \dotfill (2)$$

If every infective bite upon a healthy person leads to sickness, then the expression (2) will show directly the number of new infections per unit of time, which we can put in the second place of the first line of the equations (1). By analogous reasoning it follows that if every person is bitten on an average b times per unit of time, the total number of infective people fz will be bitten bfz times, and a fraction $\dfrac{p^1 - z^1}{p^1}$ of these bites will be made by healthy mosquitoes which will thus become infected. Consequently the number of new infections among mosquitoes per unit of time will be:

$$bfz \frac{p^1 - z^1}{p^1} \dotfill (3)$$

Now, evidently, the total number of mosquito bites on human beings per unit of time will constitute a certain fixed value, which can be written either as b^1p^1, i.e., the product of the number of mosquitoes by the number of bites made by each mosquito per unit of time, or as bp, i.e., the number of persons multiplied by the number of times each human being has been bitten. We have therefore $bp = b^1 p^1$, and finally

$$b = \frac{b^1 p^1}{p} \dots\dots\dots\dots\dots\dots\dots (4)$$

Inserting the expression (4) into the formula (3) we obtain:

$$\frac{b^1 p^1 fz}{pp^1} (p^1 - z^1) = \frac{b^1 fz}{p} (p^1 - z^1) \dots\dots\dots\dots (5)$$

The expression (5) fills the second place in the lower line in Ross's differential equations of malaria (1). It gives the number of new infections of mosquitoes per unit of time. We can now put down the rate of increase of infected individuals among the human population as $\frac{dz}{dt}$, and the rate of increase in the number of infected mosquitoes as $\frac{dz^1}{dt}$. The number of recoveries per unit of time among human individuals will be rz, as z represents the number of people infected and r the rate of recovery, i.e., the fraction of the infected population recovering per unit of time. The number of infected mosquitoes dying per unit of time can be put down as $M^1 z^1$, since z^1 denotes the number of mosquitoes infected, and M^1 the death rate in mosquitoes per head per unit of time. We can now express the equation of Ross in mathematical symbols instead of words:

$$\left.\begin{array}{l} \dfrac{dz}{dt} = b^1 f^1 z^1 \dfrac{p - z}{p} - rz \\[3mm] \dfrac{dz^1}{dt} = b^1 fz \dfrac{p^1 - z^1}{p} - M^1 z^1 \end{array}\right\} \dots\dots\dots\dots\dots (6)$$

These simultaneous differential equations of the struggle for existence express in a very simple and clear form the continuous dependence of the infection of people on the infectivity of the mosquitoes and vice versa. The increase in the number of sick persons is connected with the number of bites made by infective mosquitoes on healthy persons per unit of time, and at the same moment the increase in the number of infected mosquitoes depends upon the bites made by healthy mosquitoes on sick people. The equations (6) enable us to

investigate the change with time (t) of the number of persons infected
with malaria (z).

The equations of Ross were submitted to a detailed analysis by
Lotka ('23, '25) who in his interesting book *Elements of Physical
Biology* gave examples of some other analogous equations.   As Lotka
remarks, a close agreement of the Ross equations with reality is not
to be expected, as this equation deals with a rather idealized case:
that of a constant population both of human beings and mosquitoes.
"There is room here for further analysis along more realistic lines.
It must be admitted that this may lead to considerable mathematical
difficulties" (Lotka ('25, p. 83)).

(3) The equations of Ross point to the important fact that a mathematical formulation of the struggle for existence is a natural consequence of simple reasoning about this process, and that it is organi-
cally connected with it.   The conditions here are more favorable than
in other fields of experimental biology.   In fact if we are engaged in
a study of the influence of temperature or toxic substances on the life
processes, or if we are carrying on investigations on the ionic theory of
excitation, the quantitative method enables us to establish in most
cases only purely empirical relations and the elaboration of a rational
quantitative theory presents considerable difficulties owing to the
great complexity of the material.   Often we cannot isolate certain
factors as we should like, and we are constantly compelled to take
into account the existence of complicated and insufficiently known
systems.   This produces the well known difficulties in applying
mathematics to the problems of general physiology if we wish to go
further than to establish purely empirical relations.   As far as the
rational mathematical theory of the struggle for existence is concerned, the situation is more favorable, because we can analyze the
properties of our species grown separately in laboratory conditions,
then make various combinations and in this way can formulate
correctly the corresponding theoretical equations of the struggle for
life.

(4) Besides the interest of the equations of Ross as the first attempt
to formulate mathematically the struggle for existence, they allow
us to answer a very common objection of biologists to such equations
in general.   It is frequently pointed out that there is no sense in
searching for exact equations of competition as this process is very
inconstant, and as the slightest change in the environmental condi-

tions or in the quantities of each species can lead to the result that instead of the second species supplanting the first it is the first species itself that begins to supplant the second. As Jennings ('33) points out, there exists a strong strain of uniformitarianism in many biologists. The idea that we can observe one effect, and then the opposite, seems to them a negation of science.

In the spreading of malaria something analogous actually takes place. Ross came to the following interesting conclusion about this matter: (1) Whatever the original number of malaria cases in the locality may have been, the ultimate malaria ratio will tend to settle down to a fixed figure dependent on the number of Anophelines and the other factors—that is if these factors remain constant all the time. (2) If the number of Anophelines is sufficiently high, the ultimate malaria ratio will become fixed at some figure between 0 per cent and 100 per cent. If the number of Anophelines is low (say below 40 per person) the ultimate malaria rate will tend to zero—that is, the disease will tend to die out. All these relations Ross expressed quantitatively, and later they were worked out very elegantly by Lotka. This example shows that a change in the quantitative relations between the components can change entirely the course of the struggle for existence. Instead of an increase of the malaria infection and its approach to a certain fixed value, there may be a decrease reaching an equilibrium with a complete absence of malaria. In spite of all this there remains a certain invariable law of the struggle for existence which Ross's equations express. In this way we see what laws are to be sought in the investigation of the complex and unstable competition processes. The laws which exist here are not of the type the biologists are accustomed to deal with. These laws may be formulated in terms of certain equations of the struggle for existence. The parameters in these equations easily undergo various changes and as a result a whole range of exceedingly dissimilar processes arises.

II

(1) The material just presented enables one to form a certain idea as to what constitutes the essence of the mathematical theories of the struggle for existence. In these theories we start by formulating the dependence of one competitor on another in a verbal form, then translate this formulation into mathematical language and obtain differen-

tial equations of the struggle for existence, which enable us to draw definite conclusions about the course and the results of competition. Therefore all the value of the ulterior deductions depends on the question whether certain fundamental premises have been correctly formulated. Consequently before proceeding any further to consider more complicated mathematical equations of the struggle for existence we must with the greatest attention, relying upon the experimental data already accumulated, decide the following question: what are the premises we have a right to introduce into our differential equations? As the problem of the struggle for existence is a question of the growth of mixed populations and of the replacement of some components by others, we ought at once to examine this problem: what is exactly known about the multiplication of animals and the growth of their homogeneous populations?

Of late years among ecologists the idea has become very wide spread that the growth of homogeneous populations is a result of the interaction of two groups of factors: the biotic potential of the species and the environmental resistance [Chapman '28, '31]. The biotic potential[1] represents the potential rate of increase of the species under given conditions. It is realized if there are no restrictions of food, no toxic waste products, etc. Environmental resistance can be measured by the difference between the potential number of organisms which can appear during a fixed time in consequence of the potential rate of increase, and the actual number of organisms observed in a given microcosm at a determined time. Environmental resistance is thus expressed in terms of reduction of some potential rate of increase, characteristic for the given organisms under given conditions. This idea is a correct one and it clearly indicates the essential factors which are operating in the growth of a homogeneous population of organisms.

However, as yet among ecologists the ideas of biotic potential and of environmental resistance are not connected with any quantitative conceptions. Nevertheless Chapman in his interesting book *Animal Ecology* arrives at the conclusion that any further progress here can only be achieved on a quantitative basis, and that in future "this direction will probably be one of the most important fields of biological science, which will be highly theoretical, highly quantitative, and highly practical."

---

[1] What Chapman calls a "partial potential."

(2) There is no need to search for a quantitative expression of the potential rate of increase and of the environmental resistance, as this problem had already been solved by Verhulst in 1838 and quite independently by Raymond Pearl and Reed in 1920. However, ecologists did not connect their idea of biotic potential with these classical works. The logistic curve, discovered by Verhulst and Pearl, expresses quantitatively the idea that the growth of a population of organisms is at every moment of time determined by the relation between the potential rate of increase and "environmental resistance." The rate of multiplication or the increase of the number of organisms (N) per unit of time (t) can be expressed as $\frac{dN}{dt}$. The rate of multiplication depends first on the potential rate of multiplication of each organism (b), i.e., on the potential number of offspring which the organism can produce per unit of time. The total potential number of offspring that can be produced by all the organisms per unit of time can be expressed as the product of the number of organisms (N) and the potential increase (b) from each one of them, i.e., bN. Therefore the potential increase of the population in a certain infinitesimal unit of time will be expressed thus:

$$\frac{dN}{dt} = bN \dots\dots\dots\dots\dots\dots\dots(7)$$

This expression represents a differential equation of the population growth which would exist if all the offspring potentially possible were produced and actually living. It is an equation of geometric increase, as at every given moment the rate of growth is equal to the number of organisms (N) multiplied by a certain constant (b).

As has been already stated, the potential geometrical rate of population growth is not realized, and its reduction is due to the environmental resistance. This idea was quantitatively expressed by Pearl in such a form that the potential geometric increase at every moment of time is only partially realized, depending on how near the already accumulated size of the population (N) approaches the maximal population (K) that can exist in the given microcosm with the given level of food resources, etc. The difference between the maximally possible and the already accumulated population ($K - N$), taken in a relative form, i.e., divided by the maximal population $\left(\frac{K - N}{K}\right)$, shows the relative number of the "still vacant places" for definite species in a given microcosm at a definite moment of time. According to the number of the still vacant places only a definite part of the potential rate of increase can be realized. At the beginning of the population growth when the relative number of unoccupied places is considerable the potential increase is realized to a great extent, but when the already accumulated population approaches the maximally possible or saturating one, only an insignificant part of the biotic potential will be realized (Fig. 3). Multiplying the biotic potential of the population (bN) by the relative number of still vacant places or its "degree of realization" $\dfrac{K - N}{K}$, we shall have the increase of population per infinitesimal unit of time:

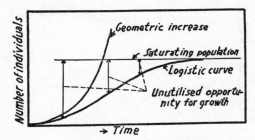

FIG. 3. The curve of geometric increase and the logistic curve

$$\begin{Bmatrix} \text{Rate of growth} \\ \text{or increase per} \\ \text{unit of time} \end{Bmatrix} = \begin{Bmatrix} \text{Potential} \\ \text{increase of} \\ \text{population} \\ \text{per unit of} \\ \text{time} \end{Bmatrix} \times \begin{Bmatrix} \text{Degree of realization} \\ \text{of the potential in-} \\ \text{crease. Depends on} \\ \text{the number of still} \\ \text{vacant places.} \end{Bmatrix} \dots (8)$$

Expressing this mathematically we have:

$$\frac{dN}{dt} = bN \frac{K - N}{K} \quad \dots\dots\dots\dots\dots\dots (9)$$

This is the differential equation of the Verhulst-Pearl logistic curve.[2]

(3) Before going further we shall examine the differential form

[2] It is to be noted that we have to do in all the cases with numbers of individuals per unit of volume or area, e.g., with population *densities* (N).

of the logistic curve in a numerical example. Let us turn our attention to the growth of a number of individuals of an infusorian, *Paramecium caudatum*, in a small test tube containing 0.5 cm³ of nutritive medium (with the sediment; see Chapter V). The technique of experimentation will be described in detail further on. Five individuals of *Paramecium* (from a pure culture) were placed in such a microcosm, and for six days the number of individuals in every tube was counted daily. The average data of 63 separate counts are given in Figure 4. This figure shows that the number of individuals in the tube increases, rapidly at first and then more slowly, until towards the fourth day it attains a certain maximal level saturating

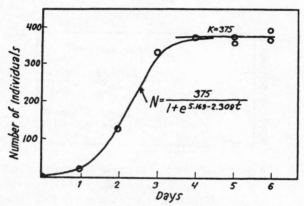

Fig. 4. The growth of population of *Paramecium caudatum*

the given microcosm. The character of the curve should be the same if we took only one mother cell at the start. Indeed, if one *Paramecium* is isolated and its products segregated as a *pure culture*, the generation time of each cell is not identically the same as that of its neighbors, and consequently at any given moment some cells are dividing, whereas the others are at various intermediate stages of the reproductive cycle; it is, however, no longer possible to divide the population up into permanent categories, since a *Paramecium* which divides rapidly tends to give rise to daughter cells which divide slowly, and vice versa. The rate of increase of such a population will be determined by the percentage of cells actually dividing at any instant, and the actual growth of the population can be plotted as a smooth curve, instead of a series of points restricted to the end of each reproductive period. The smooth curve of Figure 4 is drawn according to the equation of the logistic curve, and its close coincidence with the results of the observations shows that the logistic curve represents a good empirical description of the growth of the population. The practical method of fitting such an empirical curve will also be considered further (Appendix II). The question that interests us just now is this: what is, according to the logistic curve, the potential rate of increase of *Paramecium* under our conditions, and how does it become reduced in the process of growth as the environmental resistance increases?

According to Figure 4, the maximal possible number of Paramecia in a microcosm of our type, or the saturating population, $K = 375$ individuals. As a result of the very simple operation of fitting the logistic curve to the empirical observations, the coefficient of multiplication or the biotic potential of one *Paramecium* (b) was found. It is equal to 2.309. This means that per unit of time (one day) under our conditions of cultivation every *Paramecium* can potentially give 2.309 new Paramecia. It is understood that the coefficient b is taken from a differential equation and therefore its value automatically obtained for a time interval equal to one day is extrapolated from a consideration of infinitesimal sections of time. This value would be realized if the conditions of an unoccupied microcosm, i.e., the absence of environmental resistance existing only at the initial moment of time, existed during the entire 24 hours. It is automatically taken into account here that if at the initial moment the population increases by a certain infinitesimal quantity proportional to this population, at the next moment the population plus the increment will increase again by a certain infinitesimal quantity proportional no longer to the initial population, but to that of the preceding moment. The coefficient b represents the rate of increase in the absence of environmental resistance under certain fixed conditions. At another temperature and under other conditions of cultivation the value b will be different. Table IV gives the constants of growth of the population of Paramecia calculated on the basis of the logistic curve. There is shown N or the number of Paramecia on the first, second, third and fourth days of growth. These numbers represent the ordinates of the logistic curve which passes near the empirical observations and smoothes certain insignificant deviations. The values bN given in Table IV express the potential rate of increase of the whole population at different moments of growth, or the number of offspring which a given population of Paramecia can potentially produce within 24 hours at these moments. We must repeat here what has been already said in calculating the value b. The potential rate bN exists only within an infinitesimal time and should these conditions exist during 24 hours the values shown in Table IV would be

TABLE IV

The growth of population of Paramecium caudatum

b (coefficient of multiplication; potential progeny per individual per day) = 2.309.

K (maximal population) = 375.

	TIME IN DAYS			
	1	2	3	4
N (number of individuals according to the logistic curve)	20.4	137.2	319.0	369.0
bN (potential increase of the population per day)	47.1	316.8	736.6	852.0
$\frac{K-N}{K}$ (degree of realization of the potential increase)	0.945	0.633	0.149	0.016
$1 - \frac{K-N}{K}$ (environmental resistance)	0.055	0.367	0.851	0.984
$\frac{dN}{dt} = bN\frac{K-N}{K}$ (rate of growth of the population)	44.5	200.0	109.7	13.6
$\frac{bN - \frac{dN}{dt}}{\frac{dN}{dt}}$ (intensity of the struggle for existence)	0.058	0.584	5.72	61.7

obtained. The expression $\frac{K-N}{K}$ shows the relative number of yet unoccupied places. At the beginning of the population growth when N is very small the value $\frac{K-N}{K}$ approaches unity. In other words the potential rate of growth is almost completely realized. As the population grows, $\frac{K-N}{K}$ approaches zero. The environmental resistance can be measured by that part of the potential increase which has not been realized—the greater the resistance the larger the unrealized part. This value can be obtained by subtracting $\dfrac{K - N}{K}$ from unity. At the beginning of growth the environmental resistance is small and $1 - \dfrac{K - N}{K}$ approaches zero. As the population increases

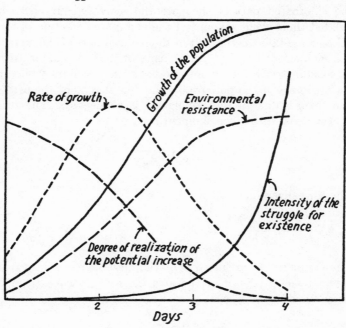

Fig. 5. The characteristics of competition in a homogeneous population of *Paramecium caudatum*.

the environmental resistance increases also and $1 - \dfrac{K - N}{K}$ approaches unity. This means that the potential increase remains almost entirely unrealized. Multiplying bN by $\dfrac{K - N}{K}$ for a given moment we obtain a rate of population growth $\dfrac{dN}{dt}$, which increases at first and then decreases. The corresponding numerical data are given in Table IV and Figure 5.

(4) We must now analyze a very important principle which was clearly understood by Darwin, but which is still waiting for its rational quantitative expression. I mean *the intensity of the struggle for existence* between individuals of a given group.[3] *The intensity of the struggle for existence is measured by the resistance which must be overcome in order to increase the number of individuals by a unit at a given moment of time.* As we measure the environmental resistance by the eliminated part of the potential increase, our idea can be formulated thus: what amount of the eliminated fraction of the potential increase falls upon a unit of the realized part of the increase at a given moment of time? The intensity of the struggle for existence keeps constant only for an infinitesimal time and its value shows with what losses of the potentially possible increment the establishment of a new unit in the population is connected. The realized value of increase at a given moment is equal to

$$\frac{dN}{dt} = bN \, \frac{K - N}{K}$$

and the unrealized one:

$$bN \left(1 - \frac{K - N}{K} \right) = bN - bN \, \frac{K - N}{K} = bN - \frac{dN}{dt} \, .$$

Then the amount of unrealized potential increase per unit of realized increase, or the intensity of the struggle for existence (i), will be expressed thus:

$$i = \frac{\begin{array}{c}\text{unrealized part of the}\\\text{potential increase}\end{array}}{\begin{array}{c}\text{realized part of the}\\\text{potential increase}\end{array}} = \frac{bN - \dfrac{dN}{dt}}{\dfrac{dN}{dt}} \quad \dots\dots\dots\dots (10)$$

[3] As we have seen in Chapter II, botanists are beginning to deal with the intensity of the struggle for existence, simply characterizing it by the per cent of destroyed individuals. Haldane investigating the connection of the intensity of competition with the intensity of selection ('31) and in his interesting book *The Causes of Evolution* ('32) specifies the intensity of competition by Z and determines it as the proportion of the number of eliminated individuals to that of the surviving ones. Thus if the mortality is equal to 9 per cent, $Z = 9/91$, i.e., approximately 0.1.

The values of i for a population of Paramecia are given in Table IV and we see that at the beginning of the population growth the intensity of the struggle for existence is not great, but that afterwards it increases considerably. Thus on the first day there are 0.058 "unrealized" Paramecia for every one realized, but on the fourth day 61.7 "unrealized" ones are lost for one realized (Fig. 6). Figure 5 shows graphically the changes of all the discussed characteristics in the course of the population growth of Paramecia.

(5) The intensity of the struggle for existence can evidently be expressed in this form only in case the population grows, i.e., if the number of individuals increases continually. If growth ceases the population is in a state of equilibrium, and the rate of growth $\frac{dN}{dt} = 0$;

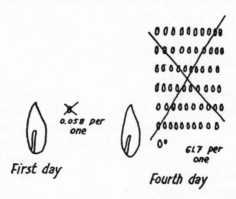

FIG. 6. Intensity of the struggle for existence in *Paramecium caudatum*. On the first day of the growth of the population 0.058 "unrealized" Paramecia are lost per one realized, but on the fourth day 61.7 per one.

in this case the expression of intensity will take another form. Population in a state of equilibrium represents a stream moving with a certain rapidity: per unit of time a definite number of individuals perishes, and new ones take their places. The number of these liberated places is not large if compared with the number of organisms that the population can produce in the same unit of time according to the potential coefficients of multiplication. Therefore a considerable part of the potentially possible increase of the population will not be realized and liberated places will be occupied only by a very small fraction of it. If the potential increase of a population in the state of equilibrium per infinitesimal unit of time is $b_1 N_1$, and a certain part (η) of this increase takes up the liberated places, it is evident that a part $(1 - \eta)$ will remain unrealized. The mechanism of this "nonrealization" is of course different in different animals. Then as before the intensity of the struggle for existence, or the proportion of the unrealized part of the potential increase to the realized part, will be:

$$i = \frac{(1 - \eta)\,b_1\,N_1}{\eta\,b_1\,N_1} = \frac{1 - \eta}{\eta} \quad \dots\dots\dots\dots (10')$$

(6) We may say that the Verhulst-Pearl logistic curve expresses quantitatively and very simply the struggle for existence which takes place between individuals of a homogeneous group. Further on we shall see how complicated the matter becomes when there is competition between individuals belonging to two different species. But one must not suppose that an intragroup competition is a very simple thing even among unicellular organisms. Though in the majority of cases the symmetrical logistic curve with which we are now concerned expresses satisfactorily the growth of a homogeneous population, certain complicating factors often appear and for some species the curves are asymmetrical, i.e., their concave and convex parts are not similar. Though this does not alter our reasoning, we have to take into account a greater number of variables which complicate the situation. Here we agree completely with the Russian biophysicist P. P. Lasareff ('23) who has expressed on this subject, but in connection with other problems, the following words: "For the development of a theory it is particularly advantageous if experimental methods and observations do not at once furnish data possessing a great degree of accuracy and in this way enable us to ignore a number of secondary accompanying phenomena which make difficult the establishment of simple quantitative laws. In this respect, e.g., the observations of Tycho Brahe which gave to Kepler the materials for formulating his laws, were in their precision just sufficient to characterize the movement of the planets round the sun to a first approximation. If on the contrary Kepler had had at his disposal the highly precise observations which we have at present, then certainly his attempt to find an empirical law owing to the complexity of the whole phenomenon could not have led him to simple and sufficiently clear results, and would not have given to Newton the material out of which the theory of universal gravitation has been elaborated.

"The position of sciences in which the methods of experimentation and the theory develop hand in hand is thus more favorable than the position of those fields of knowledge where experimental methods far outstrip the theory, as is, e.g., observed in certain domains of experimental biology, and then the development of the theory becomes more difficult and complicated" (p. 6).

Therefore we must not be afraid of the simplicity of the logistic curve for the population of unicellular organisms and criticize it from this point of view. At the present stage of our knowledge it is just sufficient for the rational construction of a theory of the struggle for existence, and the secondary accompanying circumstances investigators will discover in their later work.[4]

(7) The application of quantitative methods to experimental biology presents such difficulties, and has more than once led to such erroneous results that the reader would have the right to consider very sceptically the material of this chapter. It is very well known that the differential equations derived from the curves observed in an experiment can be only regarded as empirical expressions and they do not throw any real light on the underlying factors which control the growth of the population. The only right way to go about the investigation is, as Professor Gray ('29) says, a direct study of factors which control the growth rate of the population and the expression of these factors in a quantitative form. In this way real differential equations will be obtained and in their integrated form they will harmonize with the results obtained by observation. In our experimental work described in the next chapter special attention has been given to a direct study of the factors controlling growth in the simplest populations of yeast cells. It has become evident that *the value*

[4] The usual objection to the differential logistic equation is that it is too simple and does not reflect all the "complexity" of growth of a population of lower organisms. In Chapter V we shall see that this remark is to a certain extent true for some populations of Protozoa. The realization of the biotic potential in certain cases actually does not *gradually diminish* with the decrease in the unutilized opportunity for growth. But this ought not to frighten any one acquainted with the methodology of modern physics: it is evident that the expression $\dfrac{K - N}{K}$ is but a first approximation to what actually exists, and, if necessary, it can be easily generalized by introducing before N a certain coefficient, which would change with the growth of the culture (for further discussion see Chapter V).

*of the environmental resistance which we have determined on the basis of
a purely physiological investigation coincides completely with the value
of the environmental resistance calculated according to the logistic equation,* using the latter as an empirical expression of growth. In this
way we have proved that the logistic equation actually expresses the
mechanism of the growth of the number of unicellular organisms
within a limited microcosm. All this will be described in detail in
the next chapter.

III

(1) We are now sufficiently prepared for the acquaintance with the
mathematical equations of the struggle for existence, and for a critical
consideration of the premises implied in them. Let us consider first
of all the case of competition between two species for the possession
of a common place in the microcosm. This case was considered
theoretically for the first time by Vito Volterra in 1926. An experimental investigation of this case was made by Gause ('32b), and at
the same time Lotka ('32b) submitted it to a further analysis along
theoretical lines.

If there is competition between two species for a common place in
a limited microcosm, we can quite naturally extend the premises implied in the logistic equation. The rate of growth of each of the
competing species in a mixed population will depend on (1) the potential rate of population increase of a given species (b_1N_1 or b_2N_2)
and (2) on the unutilized opportunity for growth of this species, just
as in the case of a population of the first and second species growing
separately. But unutilized opportunity for growth of a given species
in a mixed population is a complex variable. It measures the number of places which are still vacant for the given species in spite of
the presence of another species, which is consuming the common food,
excreting waste products and thereby depriving the first one of some
of the places. Let us denote as before by N_1 the number of individuals of the first species, though, as we shall see further on, we shall
have to deal in many cases not with the numbers of individuals but
with masses of species (= the weight of the organisms present or its
equivalent) and we shall introduce corresponding alterations. The
unutilized opportunity for growth, or the degree of realization of the
potential increase for the first species in a mixed population, may be
expressed thus: $\dfrac{K_1 - (N_1 + m)}{K_1}$, where K_1 is the maximal possible

number of individuals of this species when grown separately under given conditions, N_1 is the already accumulated number of individuals of the first species at a given moment in the mixed population, and m is the number of the places of the first species in terms of the number of individuals of this species, which are taken up by the second species at a given moment. The unutilized opportunity for growth of the first species in the mixed population can be better understood if we compare it with the value of the unutilized opportunity for the separate growth of the same species. In the latter case the unutilized opportunity for growth is expressed by the difference (expressed in a relative form) between the maximal number of places and the number of places already occupied by the given species. Instead of this for the mixed population we write the difference between the maximal number of places and that of the places already taken up by *our species together with the second species* growing simultaneously.

(2) An attempt may be made to express the value m directly by the number of individuals of the second species at a given moment, which can be measured in the experiment. But it is of course unlikely that in nature two species would utilize their environment in an absolutely identical way, or in other words that equal numbers of individuals would consume (on an average) equal quantities of food and excrete equal quantities of metabolic products of the same chemical composition. Even if such cases do exist, as a rule different species do not utilize the environment in the same way. Therefore the number of individuals of the second species accumulated at a given moment of time in a mixed population in respect to the place it occupies, which might be suitable for the first species, is by no means equivalent to the same number of individuals of the first species. The individuals of the second species have taken up a certain larger or smaller place. If N_2 expresses the number of individuals of the second species in a mixed population at a given moment, than the places of the first species which they occupy *in terms of the number of individuals of the first species,* will be $m = \alpha N_2$. Thus, the coefficient α is the coefficient reducing the number of the individuals of the second species to the number of places of the first species which they occupy. This coefficient α shows the degree of influence of one species upon the unutilized opportunity for growth of another. In fact, if the interests of the different species do not clash and if in the microcosm they occupy places of a different type or different "niches" then the degree of influence of one species on the opportunity for

growth of another, or the coefficient α, will be equal to zero. But if the species lay claim to the very same "niche," and are more or less equivalent as concerns the utilization of the medium, then the coefficient α will approach unity. And finally if one of the species utilizes the environment very unproductively, i.e., if each individual consumes a great amount of food or excretes a great quantity of waste products, then it follows that an individual of this species occupies as large a place in the microcosm as would permit another species to produce many individuals, and the coefficient α will be large. In other words an individual of this species will occupy the place of many individuals of the other species. If we remember here the specificity of the metabolic products, and all the very complex relations which can exist between the species, we shall understand how useful we may find the coefficient of the struggle for existence α, which objectively shows how many places suitable for the first species are occupied by one individual of the second.

Taking the coefficient α, we can now express in the following manner the unutilized opportunity for growth of the first species in a mixed population: $\dfrac{K_1 - (N_1 + \alpha N_2)}{K_1}$. The unutilized opportunity for growth of the second species in a mixed population will have a similar expression: $\dfrac{K_2 - (N_2 + \beta N_1)}{K_2}$. The coefficient of the struggle for existence β indicates the degree of influence of every individual of the first species on the number of places suitable for the life of the second species. These two expressions enable one to judge in what degree the potential increase of each species is realized in a mixed population.

(3) As we have already mentioned in analyzing the Ross equations, an important feature of mixed populations is the simultaneous influence upon each other of the species constituting them. The rate of growth of the first species depends upon the number of places already occupied by it as well as by the second species at a given moment. As growth proceeds the first species increases the number of places already occupied, and thus affects the growth of the second species as well as its own. We can introduce the following notation:

$\dfrac{dN_1}{dt}, \dfrac{dN_2}{dt}$ = rates of growth of the number of individuals of the first and second species in a mixed population at a given moment.

N_1, N_2 = number of individuals of the first and second species in a mixed population at a given moment.

b_1, b_2 = potential coefficients of increase in the number of individuals of the first and second species.

K_1, K_2 = maximal numbers of individuals of the first and second species under the given conditions when separately grown.

α, β = coefficients of the struggle for existence.

The rate of growth of the number of individuals of the first species in a mixed population is proportional to its potential rate $(b_1 N_1)$, which in every infinitesimal time interval is realized in greater or less degree depending on the relative number of the still vacant places: $\dfrac{K_1 - (N_1 + \alpha N_2)}{K_1}$. An analogous relationship holds true for the second species. The growth of the first and second species is simultaneous. It can be expressed by the following system of simultaneous differential equations:

$$
\left.
\begin{cases}
\text{Rate of growth} \\
\text{of the first spe-} \\
\text{cies in a mixed} \\
\text{population}
\end{cases}
=
\begin{cases}
\text{Potential in-} \\
\text{crease of the} \\
\text{population of} \\
\text{the first species}
\end{cases}
\times
\begin{cases}
\text{Degree of reali-} \\
\text{zation of the po-} \\
\text{tential increase.} \\
\text{Depends on the} \\
\text{number of still} \\
\text{vacant places.}
\end{cases} \\
\begin{cases}
\text{Rate of growth} \\
\text{of the second} \\
\text{species in a} \\
\text{mixed popula-} \\
\text{tion}
\end{cases}
=
\begin{cases}
\text{Potential in-} \\
\text{crease of the} \\
\text{population of} \\
\text{the second spe-} \\
\text{cies}
\end{cases}
\times
\begin{cases}
\text{Degree of reali-} \\
\text{zation of the po-} \\
\text{tential increase.} \\
\text{Depends on the} \\
\text{number of still} \\
\text{vacant places.}
\end{cases}
\right\} \dots (11)
$$

Translating this into mathematical language we have:

$$
\left.
\begin{aligned}
\frac{dN_1}{dt} &= b_1 N_1 \frac{K_1 - (N_1 + \alpha N_2)}{K_1} \\
\frac{dN_2}{dt} &= b_2 N_2 \frac{K_2 - (N_2 + \beta N_1)}{K_2}
\end{aligned}
\right\} \dots\dots\dots\dots (12)
$$

The equations of the struggle for existence which we have written express quantitatively the process of competition between two species for the possession of a certain common place in the microcosm. They are founded on the idea that every species possesses a definite potential coefficient of multiplication but that the realization of these potentialities (b_1N_1 and b_2N_2) of two species is impeded by *four* processes hindering growth: (1) in increasing the first species diminishes its own opportunity for growth (accumulation of N_1), (2) in increasing the second species decreases the opportunity for growth of the first species (αN_2), (3) in increasing the second species decreases its own opportunity for growth (accumulation of N_2), and (4) the increase of the first species diminishes the opportunity for growth of the second species (βN_1). Whether the first species will be victorious over the second, or whether it will be displaced by the second depends, first, on the properties of each of the species taken separately, i.e., on the potential coefficients of increase in the given conditions (b_1, b_2), and on the maximal numbers of individuals (K_1, K_2). But when two species enter into contact with one another, new coefficients of the struggle for existence α and β begin to operate. They characterize the degree of influence of one species upon the growth of another, and participate in accordance with the equation (12) in producing this or that outcome of the competition.

(4) It is the place to note here that the equation (12) as it is written does not permit of any equilibrium between the competing species occupying the same "niche," and leads to the entire displacing of one of them by another. This has been pointed out by Volterra ('26), Lotka ('32b) and even earlier by Haldane ('24), and for the experimental confirmation and a further analysis of this problem the reader is referred to Chapter V. We can only remark here that this is immediately evident from the equation (12). The stationary state occurs whenever $\frac{dN_1}{dt}$ and $\frac{dN_2}{dt}$ both vanish together $\left(\frac{dN_1}{dt} = \frac{dN_2}{dt} = 0\right)$, and the mathematical considerations show that with *usual* α and β there cannot simultaneously exist positive values for both $N_{1, \infty}$ and $N_{2, \infty}$. One of the species must eventually disappear. This apparently harmonizes with the biological observations. As we have pointed out in Chapter II, both species survive indefinitely only when they occupy different niches in the microcosm in which they have an advantage over their competitors. Experimental investigations of such complicated systems are in progress at the time of this writing.

(5) We have just discussed a very important set of equations of the competition of two species for a common place in the microcosm, and it remains to make in this connection a few historical remarks. Analogous equations dealing with a more special case of competition between two species for a common food were for the first time given in 1926 by the Italian mathematician Vito Volterra who was not acquainted with the investigations of Ross and of Pearl.

Volterra assumed that the increase in the number of individuals obeys the law of geometric increase: $\frac{dN}{dt} = bN$, but as the number of individuals (N) accumulates, the coefficient of increase (b) diminishes to a first approximation proportionally to this accumulation ($b - \lambda N$), where λ is the coefficient of proportionality. Thus we obtain:

$$\frac{dN}{dt} = (b - \lambda N) N \quad \dots\dots\dots\dots\dots(13)$$

It can be easily shown, as Lotka ('32) remarks, that the equation of Volterra (13) coincides with the equation of the logistic curve of Verhulst-Pearl (9). In fact, if we call the rate of growth per individual a relative rate of growth and denote it as: $\frac{1}{N}\frac{dN}{dt}$, then the equation (13) will have the following form:

$$\frac{1}{N} \cdot \frac{dN}{dt} = b - \lambda N \quad \dots\dots\dots\dots\dots(14)$$

This enables us to formulate the equation (13) in this manner: the relative rate of growth represents a linear function of the number of individuals N, as $b - \lambda N$ is the equation of a straight line. If we now take the equation of the Verhulst-Pearl logistic curve (9): $\frac{dN}{dt} = bN \frac{K - N}{K}$, and make the following transformations:

$$\frac{dN}{dt} = bN \left(1 - \frac{1}{K} N\right); \quad \frac{1}{N} \cdot \frac{dN}{dt} = b \left(1 - \frac{1}{K} N\right), \text{ we shall have:}$$

$$\frac{1}{N} \cdot \frac{dN}{dt} = b - \frac{b}{K} N \quad \dots\dots\dots\dots\dots(15)$$

In other words the logistic curve possesses the property that with an increase in the number of individuals the relative rate of growth decreases linearly (this has been recently mentioned by Winsor ('32)). Consequently the expression (13) according to which we must subtract from the coefficient of increase b a certain value proportional to the accumulated number of individuals in order to obtain the rate of growth, and the expression (9), according to which we must multiply the geometric increase bN by a certain "degree of its realization," coincide with one another. Both are based on a broad mathematical assumption of a linear relation between the relative rate of growth and the number of individuals. Volterra extended the equation (13) to the competition of two species for common food, assuming that the presence of a certain number of individuals of the first species (N_1) decreases the quantity of food by h_1N_1, and the presence of N_2 individuals of the second species decreases the quantity of food by h_2N_2. Therefore, both species together decrease the quantity of food by $h_1N_1 + h_2N_2$, and the coefficient of multiplication of the first species decreases in connection with the diminution of food:

$$b_1 - \lambda_1 (h_1N_1 + h_2N_2) \dots\dots\dots\dots\dots (16)$$

But for the second species the degree of influence of the decrease of food on the coefficient of multiplication b_2 will be different (λ_2), and we shall obtain:

$$b_2 - \lambda_2 (h_1N_1 + h_2N_2) \dots\dots\dots\dots\dots (17)$$

Starting from these expressions Volterra ('26) wrote the following simultaneous differential equations of the competition between two species for common food:

$$\left.\begin{array}{l} \dfrac{dN_1}{dt} = [b_1 - \lambda_1(h_1 N_1 + h_2 N_2)] N_1 \\[2mm] \dfrac{dN_2}{dt} = [b_2 - \lambda_2(h_1 N_1 + h_2 N_2)] N_2 \end{array}\right\} \dots\dots\dots\dots (18)$$

These equations represent, therefore, a natural extension of the principle of the logistic curve, and the equation (12) written by Gause ('32b) coincides with them. Indeed the equation (12) can be transformed in this manner:

$$\frac{dN_1}{dt} = b_1 N_1 \left[1 - \left(\frac{1}{K_1} N_1 + \frac{\alpha}{K_1} N_2 \right) \right]$$
$$\frac{dN_2}{dt} = b_2 N_2 \left[1 - \left(\frac{1}{K_2} N_2 + \frac{\beta}{K_2} N_1 \right) \right]$$
or

$$\frac{dN_1}{dt} = \left[b_1 - b_1 \left(\frac{1}{K_1} N_1 + \frac{\alpha}{K_1} N_2 \right) \right] N_1$$
$$\frac{dN_2}{dt} = \left[b_2 - b_2 \left(\frac{1}{K_2} N_2 + \frac{\beta}{K_2} N_1 \right) \right] N_2$$
.........(19)

The result of the transformation shows that the equation (12) coincides with Volterra's equation (18), but it does not include any parameters dealing with the food consumption, and simply expresses the competition between species in terms of the growing populations themselves. As will be seen in the next chapter, the equation (12) is actually realized in the experiment.

IV

(1) In the present book our attention will be concentrated on an experimental study of the struggle for existence. In this connection we are interested only in those initial stages of mathematical researches which have already undergone an experimental verification. At the same time we are writing for biological readers and we would not encumber them by too numerous mathematical material. All this leads us to restrict ourselves to an examination of only a few fundamental equations of the struggle for existence, referring those who are interested in mathematical questions to the original investigations of Volterra, Lotka and others.

We shall now consider the second important set of equations of the struggle for existence, which deals with the destruction of one species by another. The idea of these equations is very near to those of Ross which we have already analyzed. They were given for the first time by Lotka ('20b) and independently by Volterra ('26). After the previous discussion these equations ought not to present any difficulties. Let us consider the process of the prey N_1 being devoured by another species, the predator N_2. We can put it in a general form:

$$\begin{Bmatrix} \text{Change in the} \\ \text{number of prey} \\ \text{per unit of time} \end{Bmatrix} = \begin{Bmatrix} \text{Natural increase} \\ \text{of prey per unit} \\ \text{of time} \end{Bmatrix} - \begin{Bmatrix} \text{Destruction of} \\ \text{the prey by the} \\ \text{predators per} \\ \text{unit of time} \end{Bmatrix}$$

$$\begin{Bmatrix} \text{Change in the} \\ \text{number of pre-} \\ \text{dators per unit} \\ \text{of time} \end{Bmatrix} = \begin{Bmatrix} \text{Increase in the} \\ \text{number of pre-} \\ \text{dators per unit} \\ \text{of time resulting} \\ \text{from the devour-} \\ \text{ing of the prey} \end{Bmatrix} - \begin{Bmatrix} \text{Deaths of the} \\ \text{predators per} \\ \text{unit of time} \end{Bmatrix} \quad \dots (20)$$

We can introduce here the following notation:

$\dfrac{dN_1}{dt}$ = rate of increase of the number of prey.

b_1 = coefficient of natural increase of prey (birth rate minus death rate).

$b_1 N_1$ = natural increase of the number of prey at a given moment.

$f_1(N_1, N_2)$ = the function characterizing the consumption of prey by predators per unit of time. This is the greater the larger is the number of predators (N_2) and the larger is the number of the prey themselves (N_1).

$\dfrac{dN_2}{dt}$ = rate of increase of the number of predators.

$F(N_1, N_2)$ = the function characterizing simultaneously the natality and the mortality of predators.

We can now translate the equations (20) into mathematical language by writing:

$$\left. \begin{aligned} \frac{dN_1}{dt} &= b_1 N_1 - f_1(N_1, N_2) \\ \frac{dN_2}{dt} &= F(N_1, N_2) \end{aligned} \right\} \quad \dots\dots\dots\dots (21)$$

In a particular case investigated by Volterra in detail, the functions in these equations have been somewhat simplified. He put $f_1(N_1, N_2) = k_1 N_2 N_1$, e.g., the consumption of prey by predators is directly proportional to the product of their concentrations. Also $F(N_1, N_2) =$

$k_2 N_2 N_1 - d_2 N_2$.   Here $k_2 N_2 N_1$ is the increase in the number of predators resulting from the devouring of the prey per unit of time, and $d_2 N_2$ — number of predators dying per unit of time (d_2 is the coefficient of mortality).   This translation of (20) gives

$$\left.\begin{aligned} \frac{dN_1}{dt} &= b_1 N_1 - k_1 N_2 N_1 \\ \frac{dN_2}{dt} &= k_2 N_2 N_1 - d_2 N_2 \end{aligned}\right\} \dots\dots\dots\dots (21a)$$

These equations have a very interesting property, namely the periodic solution, which has been discovered by both Lotka ('20) and Volterra ('26).   As the number of predators increases the prey diminish in number,[5] but when the concentration of the latter becomes small, the predators owing to an insufficiency of food begin to decrease.[6]   This produces an opportunity for growth of the prey, which again increases in number.

(2) In our discussion up to this point we have noted how the process of interaction between predators and prey can be expressed in a general form covering a great many special cases (equation 21), and how this general expression can be made more concrete by introducing certain simple assumptions (equation 21a).   There is no doubt that we shall not obtain any real insight into the nature of these processes by further abstract calculations, and the reader will have to wait for Chapter VI where the discussion is continued on the sound basis of experimental data.

Let us better devote the remainder of this chapter to two rather special problems of the natural increase of both predators and prey in a mixed culture simply in order to show how the biological reasoning can be translated into mathematical terms.

In the general form the rate of increase in the number of individuals of the predatory species resulting from the devouring of the prey $\dfrac{dN_2}{dt}$

---

[5] When the number of predators (N_2) is considerable the number of the prey devoured per unit of time ($k_1 N_1 N_2$) is greater than the natural increase of the prey during the same time ($+b_1 N_1$), and ($b_1 N_1 - k_1 N_1 N_2$) becomes a negative value.

[6] With a small number of prey (N_1) the increase of the predators owing to the consumption of the prey ($+k_2 N_1 N_2$) is smaller than the mortality of the predators ($-dN_2$), and ($k_2 N_1 N_2 - dN_2$) becomes a negative value.

can be represented by means of a certain geometrical increase which is realized in proportion to the unutilized opportunity of growth. This unutilized opportunity is a function of the number of prey at a given moment: $f(N_1)$. Therefore,

$$\frac{dN_2}{dt} = b_2 N_2 f(N_1) \dotso (22)$$

The simplest assumption would be that the geometric increase in the number of predators is realized in direct proportion to the number of prey (λN_1). Were our system a simple one we could say with Lotka and Volterra that the rate of growth might be directly connected with the number of encounters of the second species with the first. The number of these encounters is proportional to the number of individuals of the second species multiplied by the number of individuals of the first $(\alpha N_1 N_2)$, where α is the coefficient of proportionality. If it were so, the increase of the number of predators would be in direct proportion to the number of the prey. Indeed, if the number of the prey N_1 has doubled and is $2N_1$, the number of their encounters with the predators has also doubled, and instead of being $\alpha N_1 N_2$ is equal to $\alpha N_2 2N_1$. Consequently the increase in the number of predators instead of the former $b_2 N_2 \lambda N_1$ would become equal to $\frac{dN_2}{dt} = b_2 N_2 \lambda \, 2N_1$, and the relative increase (per predator) would be therefore: $\frac{1}{N_2} \frac{dN_2}{dt} = b_2 \lambda \, 2N_1$. In other words, the relative increase $\frac{1}{N_2} \frac{dN_2}{dt}$ would be a rectilinear function of the number of prey N_1, i.e., with a rise of the concentration of the prey the corresponding values of the relative increase of the predators could be placed on a straight line (ab in Fig. 7). But experience shows the following: If we study the influence of the increase in the number of the prey per unit of volume upon the increase from one predator per unit of time, we will find that this increase rises at first rapidly and then slowly, approaching a certain fixed value. A further change in density of the prey does not call forth any rise in the increase per predator. In the limits which interest us we can express this relationship with the aid of a curve rapidly increasing at first and then approaching a certain asymptote. Such a curve is represented in Figure 7 (ac). The concentration of prey (N_1) is marked on the abscissae, and the relative increase in the number of predators $\left(\dfrac{1}{N_2}\dfrac{dN_2}{dt}\right)$, or the rate of growth per predator at different densities of prey, is marked on the ordinates. The curve connecting the relative increase of the predators with the concentration of the prey can be expressed by the equation: $y = a\,(1 - e^{-kx})$,[7] which in our case takes the following form:

$$\frac{1}{N_2}\frac{dN_2}{dt} = b_2(1 - e^{-\lambda N_1})\dots\dots\dots\dots\dots(23)$$

The properties of this curve are such that as the prey becomes

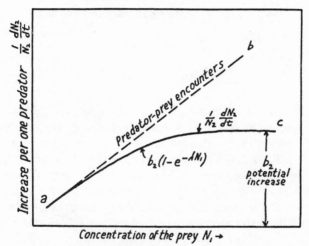

FIG. 7. The connection between the relative increase of the predators and concentration of the prey.

more concentrated the relative increase of the predators rises also, approaching gradually the greatest possible or potential increase b_2 (see Fig. 7). The meaning of the equation (23) is that instead of the assumption of a linear alteration of the relative increase which

[7] This is the simplest expression of the curve of such a type, which is widely used in modern biophysics. It is deduced from the assumption that the rate of increase is proportional to still unutilized opportunity for increase taken in an absolute form: $\dfrac{dy}{dx} = k(a - y)$. For experimental verification see Chapter VI.

was justified in the case of the competition for common food discussed before, we now take a further step and express a non-linear relation of the increase per predator with the concentration of the prey. All this will be easier to understand when in Chapter VI we pass on the analysis of the experimental material. We shall then explain also the meaning of the coefficient λ in the equation (23).

(3) The question now arises as to how to express the natural increase of the prey. As noticed already, the growth of the prey in a limited microcosm in the absence of predators can be expressed in the form of a potential geometric increase $b_1 N_1$, which at every moment of time is realized in dependence on the unutilized opportunity for growth $\dfrac{K_1 - N_1}{K_1}$. Therefore, the natural increase in the number of prey per unit of time can be expressed thus:

$$\frac{dN_1}{dt} = b_1 N_1 \frac{K_1 - N_1}{K_1} .$$

It is easy to see that in the presence of the predator devouring the prey the expression of the unutilized opportunity for growth of the latter will take a more complex form. The unutilized opportunity for growth will, as before, be expressed by the difference (taken in a relative form) between the maximal number of places which is possible under given conditions (K_1), and the number of places already occupied. But the number of prey (N_1) which in the presence of the predator exist at the given moment, does not reflect the number of the "already occupied places." In fact the prey which have been devoured by the predator have together with the actually existing prey participated in the utilization of the environment, i.e., consumed the food and excreted waste products. Therefore, the degree of utilization of the environment is determined by the total of the present population (N_1) and that which has been devoured (n).[8] The expression of the unutilized opportunity for growth of the prey will have then the following form:

[8] These calculations are true only in the case of the competition for a certain limited amount of energy (see Chapter V). In other words it is assumed that the nutritive medium is not changed in the course of the experiment. With change of the medium at short intervals (as discussed in Chapter V) the term n disappears, and the situation becomes more simple.

$$\frac{K_1 - (N_1 + n)}{K_1} \dots\dots\dots\dots\dots (25)$$

Here we arrive at a very interesting conclusion, namely, that *the development of a definite biological system is conditioned not only by its state at a given moment, but that the past history of the system exerts a powerful influence together with its present state.* This fact is apparently very widespread, and one can read about it in a general form in almost every manual of ecology. Lotka and Volterra expressed mathematically the rôle of this circumstance in the processes of the struggle for existence.

In order to complete the consideration we have to express the number of the devoured prey from the moment the predator is introduced up to the present time. If in every infinitesimal time interval the predator devours a certain definite number of prey then the total number of the devoured prey will be equal to the sum of the elementary quantities devoured from the moment the predator is introduced up to the given time (t). This total can be apparently expressed by a definite integral.

(4) We cannot at present ignore the difficulties existing in the field of the mathematical investigation of the struggle for life. We began this chapter with comparatively simple equations dealing with an idealized situation. Then we had to introduce one complication after another, and finally arrived at rather complicated expressions. But we had in view populations of unicellular organisms with an immense number of individuals, a short duration of generations and a practically uniform rate of natality. The phenomena of competition are reduced here to their simplest. What enormous difficulties we shall, therefore, encounter in attempts to find rational expressions for the growth of more complicated systems.[9] Is it worth while, on the whole, to follow this direction of investigation any further?

There is but one answer to this question. We have at present no other alternative than an analysis of the elementary processes of the struggle for life under very simple conditions. Nothing but a very active investigation will be able to decide in the future the problem of the behavior of the complicated systems. Now we can only point

---

[9] We can have an idea of this from the recent papers of Stanley ('32) and Bailey ('33) who try to formulate the equations of the struggle for existence for various insect populations.

out two principal conditions which must be realized in a mathematical investigation of the struggle for existence in order to avoid serious errors and the consequent disillusionment as to the very direction of work.   These conditions are: (1) The equations of the growth of populations must be expressed in terms of the populations themselves, i.e., in terms of the number of individuals, or rather of the biomass, constituting a definite population.   It must always be kept in mind that even in such a science as physical chemistry it is only *after* the course of chemical reactions has been quantitatively formulated in terms of the reactions themselves, that the attempt has been made to explain some of them on the ground of the kinetic theory of gases. (2) The quantitative expression of the growth of population must go hand in hand with a direct study of the factors which control growth. Only in those cases, where the results deduced from equations *are confirmed by the data obtained through entirely different methods, by a direct study of the factors limiting growth,* can we be sure of the correctness of the quantitative theories.

Chapter IV

ON THE MECHANISM OF COMPETITION IN YEAST CELLS

I

(1) No mathematical theories can be accepted by biologists without a most careful experimental verification. We can but agree with the following remarks made in *Nature* (H. T. H. P. '31) concerning the mathematical theory of the struggle for existence developed by Vito Volterra: "This work is connected with Prof. Volterra's researches on integro-differential equations and their applications to mechanics. In view of the simplifying hypothesis adopted, the results are not likely to be accepted by biologists until they have been confirmed experimentally, but this work has as yet scarcely begun." First of all, very reasonable doubts may arise whether the equations of the struggle for existence given in the preceding chapter express the essence of the processes of competition, or whether they are merely empirical expressions. Everybody remembers the attempt to study from a purely formalistic viewpoint the phenomena of heredity by calculating the likeness between ancestors and descendants. This method did not give the means of penetrating into the mechanism of the corresponding processes and was consequently entirely abandoned. In order to dissipate these doubts and to show that the above-given equations actually express the mechanism of competition, we shall now turn to an experimental analysis of a comparatively simple case. It has been possible to measure directly the factors regulating the struggle for existence in this case, and thus to verify some of the mathematical theories.

Generally speaking, biologists usually have to deal with empirical equations. The essence of such equations is admirably expressed in the following words of Raymond Pearl ('30): "The worker in practically any branch of science is more or less frequently confronted with this sort of problem: he has a series of observations in which there is clear evidence of a certain orderliness, on the one hand, and evident fluctuations from this order, on the other hand. What he obviously wishes to do . . . is to emphasize the orderliness and minimize the

59

fluctuations about it. . . . He would like an expression, exact if possible, or, failing that, approximate, of the law if there be one. This means a mathematical expression of the functional relation between the variables. . . .

"It should be made clear at the start that there is, unfortunately, no method known to mathematics which will tell anyone in advance of the trial what is either the correct or even the best mathematical function with which to graduate a particular set of data. The choice of the proper mathematical function is essentially, at its very best, only a combination of good judgment and good luck. In this realm, as in every other, good judgment depends in the main only upon extensive experience. What we call good luck in this sort of connection has also about the same basis. The experienced person in this branch of applied mathematics knows at a glance what general class of mathematical expression will take a course, when plotted, on the whole like that followed by the observations. He furthermore knows that by putting as many constants into his equation as there are observations in the data he can make his curve hit all the observed points exactly, but in so doing will have defeated the very purpose with which he started, which was to emphasize the law (if any) and minimize the fluctuations, because actually if he does what has been described he emphasizes the fluctuations and probably loses completely any chance of discovering a law.

"Of mathematical functions involving a small number of constants there are but relatively few. . . . In short, we live in a world which appears to be organized in accordance with relatively few and relatively simple mathematical functions. Which of these one will choose in starting off to fit empirically a group of observations depends fundamentally, as has been said, only on good judgment and experience. There is no higher guide" (pp. 407–408).

(2) We are now confronted by an entirely different problem which has often arisen in other domains of exact science and which represents the next step after establishing the first empirical relations without any mathematical theory. The problem is that *from clearly formulated hypotheses which appear probable on the ground of collected experimental material certain mathematical consequences are deduced, connecting the experimental values in equations accessible to experimental verification.* As a result a mathematical theory of the phenomena observed in a given field of science is obtained. The equations of the

struggle for existence are just such theoretical equations that have been deduced from hypotheses about potential coefficients of multiplication of species and the participation of these species in the utilization of a limited opportunity for growth. The verification of such a theoretical equation of the struggle for existence may be reduced to the following: (1) we must determine experimentally the potential coefficients of multiplication of the species; (2) by means of a direct study of the factors limiting growth we must evaluate the degree of influence of one species on the opportunity for growth of another, i.e., the coefficients of the struggle for existence; (3) by inserting all these values into a theoretical equation we must obtain a complete agreement with the experimental data, if our mathematical theory connects correctly the coefficients furnished by experimentation. It seems to us that these three steps of verifying our theoretical equations must be somewhat modified, taking into account the complicated situation in the competition between two species for a common place in the microcosm. We proceed as follows: (1) having determined the potential coefficients of multiplication b_1, b_2 and the maximal biomasses K_1, K_2 we pass on at once to (3), i.e., on the basis of the experimental data, taking our equations as purely empirical expressions or, in other terms, considering that they *must describe* the values observed, we calculate those empirical coefficients of the struggle for existence with which the equations *actually describe* the experimental data. It is only then that we pass to (2), and *compare these empirically found coefficients of the struggle for existence with those which are to be expected from a direct study of the factors limiting growth. If the empirical coefficients coincide with the theoretical ones, the correctness of the mathematical theory will be proved.*

This mode of verification of the mathematical theory has been adopted by us because the coincidence of theoretical coefficients with the empirical ones is but rarely to be expected. Such a rare case representing, most likely, rather an exception than a rule is described in this chapter. This small probability of a coincidence of the coefficients is connected with the fact that usually the growth of populations depends on numerous factors, many of which (e.g., waste-products) we often cannot specify exactly, and the influence of one species on the opportunity of growth of another under these conditions is realized in a very complicated manner. Hence the empirical coefficients of the struggle for existence, calculated by an equation which in certain cases has already been verified, can serve as a guide for the study of the very mechanism of the influence of one species on the growth of another.

II

(1) To verify our differential equations of the struggle for existence we had recourse to populations of yeast cells. Yeast cells were cultivated in a liquid nutritive medium, where they were nourished by various substances dissolved in water and excreted certain waste-products into the surrounding medium. Owing to the considerable practical importance of yeast for the food industry a great number of papers has been devoted to investigation of its growth, and although the majority deals with purely practical questions that do not at present interest us, nevertheless it is pretty well ascertained what substances yeast requires for its growth, and what is the chemical composition of the waste-products it excretes.

For the study of competition we took two species of yeast: (1) a pure line of common yeast, *Saccharomyces cerevisiae* stock XII, received from the Berliner Gährungsinstitut, and (2) a pure line of the yeast *Schizosaccharomyces kephir*, cultivated in the Moscow Institute of the Alcohol Industry and obtained from Dr. Pervozvansky.[1] Both these species can grow under anaerobic conditions as well as when oxygen is accessible. It is very well known that the processes of life activity are connected with a continuous consumption of energy which is supplied by certain chemical reactions. In the case when the growth of yeast proceeds in the absence of oxygen it is the decomposition of sugar into alcohol and carbon dioxide which fur-

[1] We began our experiments with yeast in 1930. The first group of experiments on competition between species was made in September–December, 1931, and appeared in the *Journal of Experimental Biology* (Gause, '32b). These experiments were extended and repeated in September–December, 1932. Their results coincided completely with the data of 1931. Later it appeared that the yeast culture kept in the Museum of the Institute of the Alcohol Industry under the name of *"Schizosaccharomyces kephir"* and used under the same name in our experiments, has been incorrectly determined by the specialists of the Museum and that it belonged to another species. The culture consists of oval, budding yeast cells much more minute than *Saccharomyces cerevisiae* and producing an alcoholic fermentation. An exact systematic determination presented extreme difficulty and seemed not to be indispensable, as this culture is kept in the Museum and can be obtained thence under the name of *"Schizosaccharomyces kephir."*

nishes the available energy, and in the nutritive medium there takes place a considerable accumulation of the waste product—ethyl alcohol.   If we alter the conditions of cultivation and allow a direct access of oxygen to the growing yeast cells, although fermentation will still continue, a part of the available energy (different for different species) will be furnished by oxidation of sugar into carbon dioxide.   In the commercial utilization of yeast, when it is desirable to accumulate alcohol in the culture, yeast is grown nearly without oxygen.   But if alcohol is not needed and the object is to obtain a great quantity of yeast cells themselves, an intensive aeration of the growing culture is carried on, which leads to an enormous increase of oxidation processes.   The yeast *Saccharomyces cerevisiae* as well as *Schizosaccharomyces kephir* produces alcoholic fermentation, and both can obtain a part of the available energy by oxidation, but they differ from one another in the relative intensities of the oxidation and fermentation processes.   Common yeast, *Saccharomyces cerevisiae*, develops well in the absence of oxygen as for it fermentation is a powerful source of energy.   It continues mainly to ferment even in the presence of oxygen (when cultivated in Erlenmeyer flasks without aeration) and utilizes the oxidation process only to a very small extent.   As regards our species of *Schizosaccharomyces*, it grows very slowly under anaerobic conditions.   However, when oxygen is available it has recourse to this source of energy; its rapidity of growth increases and it approaches *Saccharomyces* in its properties.   Hence, *Saccharomyces* represents a species with distinctly expressed fermentative capacities, whilst *Schizosaccharomyces* is a species of a more oxidizing type.   By mixing these species we obtain a very interesting situation for studying the competition between species in different conditions of environment.

(2) We cultivated yeast in a sterilized nutritive medium which was prepared in the following manner: 20 gr. of dry pressed beer-yeast were mixed with 1 liter of distilled water, boiled for half an hour in a Kochs boiler, and then filtered through infusorial earth.   Five per cent of sugar was added to this mixture, and then the medium was sterilized in an autoclave.   A medium of such a type is very favorable for the growth of yeast, because the decoction contains all the nutritive substances required.   The only disadvantage is our ignorance of the exact chemical composition of this medium.   Therefore each series of experiments must be made with a solution of the very same

preparation.   But on the whole this method enables one to have
sufficiently standardized conditions for cultivation.

The nutritive medium was sterilized in a large flask and then asep-
tically poured into small vessels for cultivation.   These vessels were
previously sterilized by dry heat (by heating to 180° for three hours).
This method has many advantages as compared with the direct
sterilization of the nutritive medium in small culture vessels.   The
fact is that when a liquid is heated in glass vessels in an autoclave,
even if the best kind of glass be used, the latter can somewhat alter
the composition of the nutritive liquid.   This produces a considerable

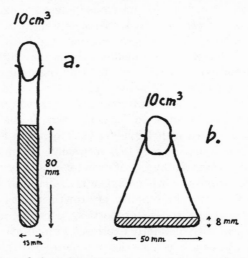

FIG. 8. The vessels for cultivation of yeast: (a) test tube, (b) Erlenmeyer's
flask.

variation in the initial conditions of separate microcosms.   The vessels used for cultivation belonged to two types: (1) in experiments
with the deficiency in oxygen we used common test tubes with a
diameter of 13 mm.   Ten cm³ of nutritive medium were poured into
such a tube, the depth of the liquid being about 80 mm.   (2) To
obtain better aeration, cultures were made in small Erlenmeyer
flasks of about 50 mm in diameter, and when 10 cm³ of nutritive
medium were poured in, the liquid reached a depth of 7–8 mm.   In
these conditions the layer of the liquid was almost ten times thinner
than in the test tubes (Fig. 8).   The test tubes as well as Erlenmeyer

flasks were closed by cotton wool stoppers. The experiments made in the flasks will be described in this book as "aerobic" and those in test tubes as "anaerobic."

(3) An inoculation of yeast cells was made into the sterilized nutritive medium. Special attention was given to the standardization of the inoculating material, for in order to obtain exact and comparable results the inoculating cells had to be in a certain fixed physiological condition. Cells for inoculation were always taken from test tubes where the growth was just finished. For an anaerobic inoculation of *Saccharomyces* cultures 48 hours old (at 28°C.) were used, whilst the slow-growing *Schizosaccharomyces* for an anaerobic inoculation was taken at the age of five days at 28°C. Before inoculation the contents of the test tube was shaken, and a fixed number of drops of the liquid was introduced into the nutritive medium by means of a sterilized pipette. It was also necessary that an equal initial quantity of each species or, in other words, equal initial masses should be inoculated. It was found that in anaerobic test tubes intended for inoculation a mass of yeast in a unit of volume of the nutritive liquid is two and a half times smaller in *Schizosaccharomyces* than in *Saccharomyces*. Therefore in order to inoculate an equal initial quantity two drops of uniform suspension of *Saccharomyces* and five drops of *Schizosaccharomyces* were always introduced. In the case of a mixed culture, two drops of the first species plus five drops of the second were taken.[2] We must prepare a perfectly uniform suspension of seed-yeast and the inoculation itself must be carried out rapidly so as to avoid possible errors from a settling of yeast cells in the inoculating pipette. This circumstance was pointed out by Richards ('32) and Klem ('33). All the experiments were carried out in a thermostat at a temperature of 28°C.

(4) After inoculation it was necessary to study the growth of number and mass of yeast cells, and on the other hand to trace and to evaluate the changes in the factors of the medium. The counting of the number of yeast cells per unit of volume does not present any difficulty and for this purpose the Thoma counting chamber is usually employed. In our experiments three test tubes (or flasks) of the

---

[2] A very strict equality of the masses of two species sown is not absolutely necessary. It is only important that the very same quantity of each species should be introduced into the mixed population and into the separately grown culture. This is very easy to do with our mode of inoculation.

same age were taken and a uniform suspension of yeast was made by shaking. One cm³ of liquid was taken by a pipette from every tube and poured into another clean tube, where the three cm³ obtained from three tubes were fixed by three cm³ of 20 per cent solution of H_2SO_4. Individual fluctuations of separate cultures were thus neutralized, and a certain "average suspension" from three test tubes was obtained. The material fixed was more or less diluted with water, and then the number of cells per unit of volume was counted in the Thoma chamber. Quite recently Richards ('32) in his interesting paper describes in detail the methods of studying the growth of yeast, where he points out that the counting of the number of yeast cells is a very satisfactory method. As regards the possible sources of error, he indicates the following: (1) the sample placed in the counting chamber is not truly representative of the population sampled; (2) the cells do not settle evenly in the counting chamber. To eliminate these errors it is necessary to take several sample groups from the "average suspension," and to count a great number of squares in the chamber. In our experiments the fixed suspension was carefully mixed before the taking of the sample, a few drops were taken with a pipette, placed in the chamber, and ten squares were counted. Six such sample groups were successively taken, and the total number of counted squares amounted to sixty. Sometimes a lesser number of squares sufficed.

The average number of cells in one large square of a Thoma chamber at the dilution corresponding to the material fixed (i.e., twice thinner than the initial suspension) is given in our tables. It is understood that the counts sometimes were made with considerably stronger dilutions, and they were correspondingly reduced to the accepted standard. A few words must be added concerning the counting of cells in mixed cultures. After a certain amount of practice it is quite easy to distinguish the two species of yeast, as the cells of *Saccharomyces* are much larger than those of *Schizosaccharomyces* and their structure is different.

(5) The numbers of yeast cells belonging to two different species do not allow us to form an idea as to their masses. But it is just the masses of the species that are of particular importance in the processes of the struggle for life. This is because a unit of mass of a given species is usually connected by definite relations with the amount of food consumed or that of the waste-products excreted or,

generally speaking, with the factors limiting growth. *Therefore the equations of the struggle for existence ought to be expressed in terms of masses of the species concerned* and not in terms of the numbers of individuals, which are connected by more complex relations with the factors limiting growth.

In order to pass on from the number of yeast cells of the first and second species counted at a definite moment to the masses of these species, we must take into account that: (1) the cells of the first species differ in their average volume from those of the second, (2) this average volume of the cell in each species can change in the course of growth of the culture. (Richards ('28b) showed that the average size of a cell of *Saccharomyces cerevisiae* is different at different stages of growth), and (3) the species can be of different specific weight. Therefore, by multiplying the volume of all the cells of a definite species at a given moment of time by their specific weight, we shall obtain the weight of the given organisms enabling us to judge of their mass. Assuming for the sake of simplification that the cells of our yeast species are near to one another in their specific weight, we can measure the volumes occupied by each species of yeast cells in order to obtain an idea of the masses of these cells.

(6) The volume of yeast was determined by the method of centrifugation. The fluid from the test tubes or flasks with the counted number of yeast cells was centrifuged for one minute in a special tube placed in an electric centrifuge making 4000 revolutions per minute (usually in portions of 10 cm³ each). The liquid was then poured off and the yeast cells that had settled on the bottom were shaken up with the small quantity of the remaining liquid. The mixture thus obtained was transferred by means of a pipette into a short graduate glass tube of 3.5 mm in diameter. The mixture in the graduated tube was again centrifuged for 1.5 minutes, and then the volume of the sediment was rapidly measured with the aid of a magnifying glass. To avoid errors connected with the different degree of compression of the yeast in different cases, the quantity of the mixture poured into the short graduated tube was always such that the sediment did not exceed ten divisions of the graduated tube and, if necessary, the secondary centrifugation was made by several doses. The volume of yeast occupying one division of the graduated tube was taken for a unit.

The centrifugation method may be criticized as, according to

Richards ('32), even in employing the super-centrifuge of Harvey one can not succeed in obtaining a solid packing of the cells, and interstices remain between them. If we draw our attention to the fact that the size of the cells changes in the process of the growth of the culture, and that in mixed populations of the two species we have to deal with cells of different sizes then, theoretically, this must lead to a very different degree of packing of the cells in different cases, and the volume of the cells determined by centrifugation apparently does not yet allow us to judge of their mass. However, the measurements, some of which will be given further on, show that the errors which actually arise are small, and that the centrifugation method is perfectly reliable for our purposes.

In the study of the population growth of yeast it is difficult to carry on observations upon the very same culture, as it is urgent to strictly maintain the sterility of the medium and to avoid injury to the cells. For this reason a great number of test tubes were inoculated at the beginning of the experiment; at certain fixed moments determinations were made upon a group of test tubes which were then put aside and further determinations were made upon new tubes.

III

(1) Having examined the technical details of cultivation of yeast cells we can now pass to the problem which interests us first of all: how does the multiplication of the yeast proceed in a microcosm with a limited amount of energy, and what are the factors which check the growth of the population? Let us begin by examining the kinetics of growth under anaerobic conditions. Figure 9 represents the growth of volume of the yeast *Saccharomyces cerevisiae*, according to the data of one of our experiments in 1930. It is clearly seen that the volume increases slowly at first, then faster, and finally slows down on approaching a certain fixed value. The curve of growth is asymmetrical, i.e., its concave part does not represent a reverse reflection of the convex one (Richards, '28c, Gause, '32a). The first of them is somewhat steep but the second comparatively inclined. This asymmetry is, however, not sharply expressed, and it can be neglected if we analyze the growth in a first approximation to reality.

In experiments of this type immediately after the yeast cells are inoculated an intensive multiplication begins. There is scarcely any lag-period, or period of an extremely slow initial growth, while the cells adapt themselves to the medium. This is because we used for inoculation fresh yeast cells developed in a medium of an identical composition with those used in the experiment. This circumstance has been pointed out by Richards ('32).

(2) An investigation of the shape of the curve which represents the accumulation of the yeast volume in the population of yeast cells does not enable us to judge what factors control the growth of the population and limit the accumulation of the biomass. The fact that the growth curve is S-shaped and resembles the well-known auto-catalytic curve does not prove at all that the phenomenon we are

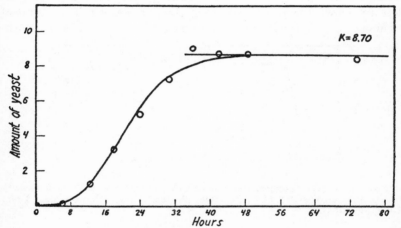

FIG. 9. Growth in volume of the yeast, *Saccharomyces cerevisiae*. From Gause ('32a).

studying has anything in common with autocatalysis. The question of the basic nature of the yeast growth in a limited microcosm can be elucidated only by means of specially arranged experiments. Such experiments were recently carried out by Richards ('28a) and confirmed by Klem ('33).

We have already mentioned that the process of multiplication of organisms is potentially unlimited. It follows the law of geometric increase, and limitations are here introduced only by the external forces. In the case of yeast this circumstance was noted by Slator ('13), and recently Richards carefully verified it in the following manner. A control culture after the inoculation of yeast was left to itself, and the growth of the number of cells in this culture followed a common S-shaped curve and then stopped. In an experimental culture a change of the medium was made at very short intervals of time (every 3 hours). Here the conditions were all the time maintained constant and favorable for growth. Under these conditions the multiplication of yeast followed the law of geometric increase: in every moment of time the increase of the population constituted a certain definite portion of the size of the population. The relative rate of growth (i.e., the rate of growth per unit of population) remained constant all the time, or in other words there was no autocatalysis here. Figure 10 represents the data of Richards. To the left are shown the growth curves of the number of cells per unit of volume: the S-shaped curve in the control culture, and the exponen-

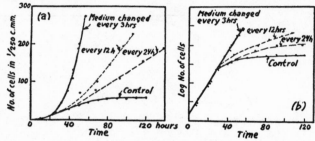

Fig. 10. Growth curves of the yeast *Saccharomyces cerevisiae*. (a) Growth of the number of cells. (b) The same, plotted on logarithmic scale. From Richards ('28a).

tially increasing one with continuously renewed medium. One can in the following manner be easily convinced that the exponentially increasing curve corresponds to the geometric increase: if against the absolute values of time we plot the logarithms of cell numbers, a straight line will be obtained (see the right part of Figure 10 taken from Richards). As is well known this is a characteristic property of a geometric increase. Nearly the same results were recently obtained by Klem ('33).

The experiments made by Richards show clearly that the growth of the yeast population is founded on a potential geometric multiplication of yeast cells ($b_1 N_1$), but the latter can not be completely realized owing to the limited dimensions of the microcosm and consequently to the limited number of places (K). As a result the geometric increase becomes S-shaped. It is easy to see that the experimentation has led us to the very same assumptions that are at the bottom of Pearl's logistic equation of growth (see Chapter III, equations (8) and (9)). This equation is one that gives us the S-shaped curve starting from the point that growth depends on a certain potential geometric increase which at every moment of time is realized only in a certain degree depending on the unutilized opportunity for growth at that moment.

In the equation of Pearl the unutilized opportunity for growth is expressed in terms of the population itself, i.e., as the relative number of the still vacant places. This presents a great advantage as we shall see later on. The unutilized opportunity of growth often depends on various factors, and to translate the number of "still vacant places" into the language of these factors may become a very difficult task.

(3) Let us now analyze this problem. What is the nature of those factors of the environment which depress the growth of the yeast population and finally stop it? Of course they may be different in various cases, and we have in view only our conditions of cultivation. The nature of the factors limiting growth in such an environment has been explained mainly by the investigations of Richards. When the growth of yeast ceases in a test tube under almost anaerobic conditions, there still exists in the nutritive medium a considerable amount of sugar and other substances necessary for growth. A simple experiment made by Richards ('28a) is convincing: if at the moment when the growth ceases in the microcosm yeast cells from young cultures are introduced, they will give a certain increment and the population will somewhat increase. Consequently, there is no lack of substances required for growth. The presence of a considerable quantity of sugar at the moment when the growth ceases has been chemically established, and in our experiments this is even more apparent than in those of Richards, as our initial concentration of sugar was 5 per cent and his only 2 per cent.

If the growth ceases before the reserves of food and energy have been exhausted we must evidently seek an explanation in some kind of changes in the environment. This question has been studied by Richards and led him to conclude that the decisive influence here is the accumulation of ethyl alcohol. As has already been mentioned, when yeast cells grow in test tubes under almost anaerobic conditions the decomposition of sugar into alcohol and carbon dioxide serves them as a source of energy. Sugar is almost entirely utilized to obtain the available energy, and serves as food only in a very slight degree. As a result a considerable amount of alcohol accumulates in the nutritive medium, which corresponds pretty well to the amount of sugar consumed. Curves of such an accumulation of alcohol,

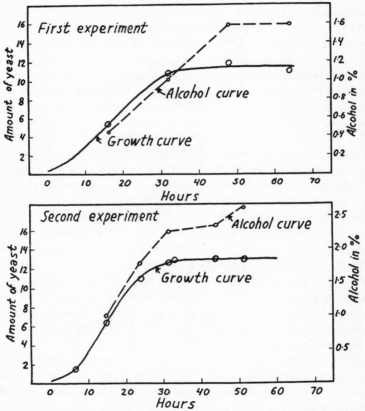

FIG. 11. The growth in volume and accumulation of alcohol in *Saccharomyces cerevisiae* in test tubes. From Gause ('32b).

taken from the paper of Gause ('32b), are represented in Figure 11. Here are given the results of two experiments made in test tubes, but on a nutritive medium of somewhat different concentration. In both cases a certain time after the experiment was begun the accumulation of alcohol (and, consequently, the consumption of sugar) proceeds almost in proportion to the increase of the volume of yeast.   In other terms, a proportionality exists between the metabolism of the yeast cells and the growth of their volume.   Later on, conditions arise in which the growth of the yeast ceases, but alcohol continues to accumulate.   Therefore, at the moment when the growth ceases there are still unutilized resources of sugar in the medium.   The life activity of the yeast cells and the accumulation of alcohol continue after the biomass has ceased growing.

The microscopical study of the population of yeast cells made by Richards at the moment when growth was ceasing, has shown the following facts.   The yeast cells continue to bud actively, but as soon

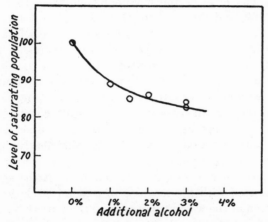

FIG. 12. The effect of additional alcohol upon the level of saturating population in *Saccharomyces cerevisiae* in test tubes.

as a bud separates from the mother cell it perishes.   In this way, unfavorable chemical changes in the medium destroy the most sensitive link in the population, and lead to a cessation of its growth. According to Richards ('28a) the accumulating ethyl alcohol is just the factor which kills the young buds and inhibits the growth of the population.   He showed this experimentally: with an addition of 1.2 per cent of ethyl alcohol to the nutritive medium, the maximal yield of population was 65 per cent from that of the control population (acidity kept constant).   Therefore, with the additional alcohol the critical concentration of waste products at which growth ceases was reached with a smaller quantity of accumulated yeast volume.

These data were criticized by Klem ('33) who carried out experiments with wort and not with William's synthetic medium, which Richards worked upon. Klem did not obtain any depression of growth by adding a small quantity of alcohol corresponding to the quantity which is usually accumulated in his cultures at the moment when the growth ceases. According to Klem, it is only at a concentration above 3 per cent that alcohol begins to depress growth, and only concentrations of about 7 per cent have a distinctly hindering influence. The experiments which I have made with yeast decoction and 5 per cent sugar confirm the data of Richards and not those of Klem. Figure 12 presents the results of several experiments. The level of the maximal population in the control was taken as 100, and the levels of the maximal populations in the cultures with this or that per cent of alcohol (added before the yeast was sown, all other conditions being equal) were expressed in per cent from the population level in the control. This figure shows that even 1 per cent of alcohol in our conditions lowers the maximal level of population considerably. As we have already seen (Fig. 11, bottom) at the moment the growth ceases in our cultures the concentration of alcohol is near to 2 per cent (with the usual composition of medium). This concentration is undoubtedly sufficiently high to be responsible for the cessation of growth.

Klem expressed an interesting idea, namely that the cessation of growth is connected with the reaching of a definite relation between the concentration of the waste-products and the nutritive substances, i.e., alcohol and sugar. In other terms, the critical concentration of alcohol checking growth is by no means of an absolute character. With a small concentration of sugar, a comparatively weak concentration of alcohol hinders growth. But if the quantity of sugar be increased, this concentration of alcohol will no longer be sufficient for checking growth which will continue. Klem's opinion is perfectly justified and many experimental data confirm it. But, as he himself remarks, the ratio alcohol/sugar left at the moment growth ceases, also varies within rather wide limits. (A critical analysis of Figs. 53–54 on pp. 80–81 of his paper ('33) shows that even with concentrations of sugar from 1 to 5 per cent the ratio alcohol/sugar left does not remain constant, and that Klem's calculations are not quite exact.)

(4) All we have said may be resumed thus: under our conditions

of cultivation the cessation of growth of the population of yeast cells begins before the exhaustion of the nutritive and energetic resources of the medium. The direct cause of this cessation is the accumulation of ethyl alcohol which kills the most sensitive members of the population—the young buds. This critical concentration of alcohol is not of an absolute character, and in a first approximation we can say that the cessation of growth is connected with the establishment of a definite ratio between the concentrations of waste-products (alcohol) and the nutritive substances (sugar). We now have to answer the question raised earlier: what factors will furnish us with the terminology for expressing the "number of vacant places" or "the unutilized opportunity for growth" in the population of yeast cells under our conditions of cultivation? Since the growth of population ceases with the establishment of a certain ratio alcohol/sugar a thought might appear that we ought to connect the unutilized opportunity for growth somehow with the ratio. However this would be a false deduction from correct premises. We can see at once that we have to deal here with two different things. (1) Should we wish to make a purely theoretical calculation of the level of saturating population in our microcosm, we would certainly be obliged to take into consideration the ratio between the concentrations of alcohol and sugar, and to try to calculate the moment when this ratio attains a definite value. But certainly we should at once have to introduce numerous corrections, as various other factors have also an influence here. (2) The conditions of the problem before us are quite different. *We know beforehand* at what level the population ceases to grow, and what is the corresponding value of different factors of the environment. We wish only for different moments of time preceding the cessation of growth to translate "the unutilized opportunity for growth" into terms of the limiting factor. Such limiting factor is always alcohol destroying the young buds. However considerably other factors of the environment and the condition of the cells themselves should alter the absolute value of the critical alcohol concentration, this does not essentially change the matter. Consequently *"the unutilized opportunity for growth" or "the number of still vacant places" can simply be determined by the difference between the critical concentration of alcohol at the moment of cessation of growth, which is characteristic for the given conditions and established experimentally in every case, and the concentration of alcohol at a given moment of time.*

The accumulation of the yeast volume at the moment of the cessation of growth is everywhere marked by K, and the amount of volume at a given moment is N. Alcohol production per unit of yeast volume is rather constant, and increases somewhat only before growth is checked (see Fig. 11). Taking the alcohol production per unit of yeast volume as a constant for the entire process of growth of the population as the very first approximation to reality, we can easily pass from the given (N) and maximal (K) amount of yeast, through multiplying them by certain coefficients, to the given and critical concentrations of alcohol.

(5) It is easy to see that, while we give up any attempt to discover a certain universal growth equation forecasting the level of the saturating population under any conditions, if we use the logistic equation we express rationally, very simply, and in complete agreement with experimental data, the mechanism of growth of a homogeneous population of yeast cells. The attempts to find universal equations will scarcely lead to satisfactory results, and in any case all this would be too complicated for a mathematical theory of the struggle for existence in a mixed population of two species. One of the leading ideas of this book is that all the quantitative theories of population growth must be only constructed for strictly determined cycles or epochs of growth, within which the same limiting factors dominate and a certain regulating mechanism remains invariable.

Experiments with yeast point also to a very important circumstance in the experimental analysis of populations. All the conditions of cultivation ought to be so arranged that the growth depends distinctly on only one limiting factor. In the case of yeast we must have a sufficiently high concentration of sugar and other necessary substances in the nutritive medium so that the alcohol can in full measure manifest its inhibitory action. As we shall see in the next chapter in experimenting with Protozoa, it is very easy to arrange experiments under such complicated conditions and with the interference of such a great number of various factors that the attempts to discover certain fundamental quantitative relations in the struggle for existence will never have any success.

IV

(1) Our study of the growth of homogeneous populations of yeast cells was only a preparation before we pass on to the investigation of

the struggle for existence between two species in a mixed culture. The simplest way to do this is again to begin by an analysis of the kinetics of growth. Let us examine the experiments of 1931. In Table 1 (Appendix) data are given on the anaerobic growth of the volume and of the number of cells in the two species of yeast: *Saccharomyces* and *Schizosaccharomyces*, cultivated separately and in a mixed population in two independent series of experiments. One hundred and eleven separate microcosms were studied in these two series, and every figure in Table 1 (Appendix) is founded on three observations. Figure 13 represents graphically the growth of the yeast volume. We can see that the growth of *Schizosaccharomyces* under anaerobic conditions is exceedingly slow. Let us note also that its population attains a much lower level than that of *Saccharo-*

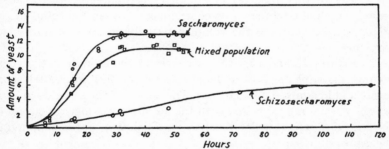

Fig. 13. The growth in volume of *Saccharomyces cerevisiae, Schizosaccharomyces kephir* and mixed population in two series of experiments. Anaerobic conditions. From Gause ('32b).

myces. The volume of the mixed population is also smaller than the volume of the pure culture of *Saccharomyces*.

The parts taken up by each of the species in the yeast volume of a mixed culture have been evaluated in the following manner. First of all, a calculation was made of the average number of cells per unit of yeast volume for the separate growth of *Saccharomyces* and *Schizosaccharomyces* (see Appendix, Table 1). It appears that the mean number of cells occupying a unit of yeast volume varies in the course of the growth of the culture, as Richards has already established. However, these variations are not great, and for further calculations average values for the entire cycle of growth can be taken. According to the first series of experiments, in *Saccharomyces* 16.59 cells in a square of a Thoma counting chamber correspond to one unit of yeast volume; in the smaller species *Schizosaccharomyces* there are 57.70 cells in one unit of yeast volume. Starting from these averages, we have calculated the volumes occupied by each species in the mixed population at a given moment, according to the number of cells of each species observed in the mixed population. (In the experiments of 1932 which are given further on we did not use such general averages for our calculations, but started every time from the average number of cells observed at a given moment of time.)

The sum of the calculated volumes of both species in the mixed culture at a given moment should agree with the actual volume of mixed population at this moment determined by the method of centrifugation. In the first series the totals of calculated volumes are somewhat smaller than the volumes actually observed, and we know the causes of this disagreement. In the second series these causes have been eliminated, and the coincidence between the totals of the volumes calculated and the volumes actually observed is a satisfactory one.

(2) Figures 14 and 15 give the curves of the growth of the yeast volume in *Saccharomyces* and *Schizosaccharomyces* cultivated separately and in a mixed population. The curves of the separate growth of each species are expressed with the aid of simple logistic curves of the following type (the details of these calculations are to be found in the Appendix):

$$\frac{dN}{dt} = bN \frac{K - N}{K},$$

where N is yeast volume, t is time, b and K are constants. The fitting of the logistic curves has given us the following values of the parameters for the separate growth of our species (Sp. No. 1 is *Saccharomyces*, No. 2 is *Schizosaccharomyces*):

Maximal volumes: $K_1 = 13.0; K_2 = 5.8$

Coefficients of geometric increase:

$$b_1 = 0.21827; \qquad b_2 = 0.06069$$

The calculated coefficients of geometric increase show that per unit of time (one hour) every unit of volume of *Saccharomyces* can potentially give an increase equal to 0.21827 of this unit, and in *Schizosaccharomyces* equal to only 0.06069.

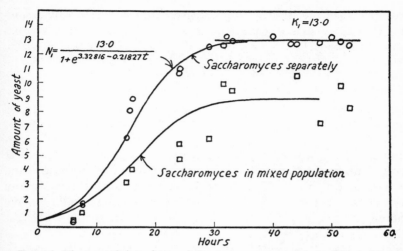

Fig. 14. The growth in volume of *Saccharomyces cerevisiae* cultivated separately and in the mixed population in two series of experiments. Anaerobic conditions. From Gause ('32b).

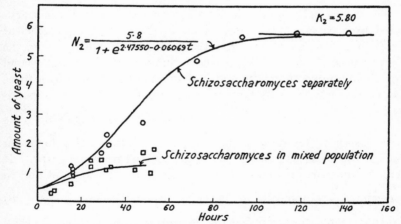

Fig. 15. The growth in volume of *Schizosaccharomyces kephir* cultivated separately and in the mixed population in two series of experiments. Anaerobic conditions. From Gause ('32b).

Having obtained in this way the potential coefficients of multiplication of our species (or, which means the same, the coefficients of geometric increase) we must now according to the general plan given at the beginning of this chapter pass on to a calculation of the empirical coefficients of the struggle for existence. In this we start by assuming that the system of equations of competition (see Chap. 3, equations (11) and (12)):

$$\left.\begin{aligned}
\frac{dN_1}{dt} &= b_1 N_1 \frac{K_1 - (N_1 + \alpha N_2)}{K_1} \\
\frac{dN_2}{dt} &= b_2 N_2 \frac{K_2 - (N_2 + \beta N_1)}{K_2}
\end{aligned}\right\}$$

actually describes the experimental data. All the values in these equations except the coefficients of the struggle for existence α and β, are known to us. To find the latter let us solve this system of two equations with two unknown values in respect to α and β. We obtain:

$$\alpha = \frac{K_1 - \dfrac{dN_1/dt \cdot K_1}{b_1 N_1} - N_1}{N_2} \; ; \qquad \beta = \frac{K_2 - \dfrac{dN_2/dt \cdot K_2}{b_2 N_2} - N_2}{N_1}$$

The values on the right side of both expressions can easily be calculated from experimental data. Thus in the case of the coefficient α: (1) b_1 and K_1 are known from the curve of separate growth of the first species, (2) N_1 and N_2, or the volumes of the first and second species in a mixed population at a given moment of time (t), can be taken from the graph by measuring the ordinates of the corresponding curves of growth, (3) $\frac{dN_1}{dt}$ represents the rate of growth of the first species in the mixed population, or the increase of volume per unit of time, and can also be easily determined from the graph. It will be sufficient for this to draw a tangent at a given point and to measure $\frac{dN_1}{dt}$ graphically or, better, to use a Richards-Roope ('30) tangent meter for graphical differentiation.[3] As a result we shall obtain the values of the coefficients of the struggle for existence (α and β) *for different points of the curve,* i.e., for different moments of growth: t_1, t_2, etc. The values of the coefficients calculated for different moments are subject to fluctuations, but by using the middle zone of growth sufficiently constant values will be obtained. Thus,

---

[3] Made by Bausch and Lomb Optical Co.

the coefficient β in the experiments of 1931 was equal to: 0.501, 0.349, 0.467, with an average of 0.439. The fluctuations of the coefficient α were more considerable, but the experiments of 1932 give more constant values for α also: 3.11, 3.06, 2.85, etc.

The fluctuations in the values of the coefficients of the struggle for existence are due in this case in a considerable measure to an imperfect method of their calculation.[4] However, this is of no serious consequence, as we have a good method for verifying the average values of the coefficients of competition. This method consists in constructing a curve corresponding to the differential equation of competition (the details of this calculation are to be found in the Appendix). A close agreement of the calculated curve of growth of each species in a mixed population with experimental observations represents a good proof of the correctness of the numerical values of the coefficients of the struggle for existence. As regards the yeasts *Saccharomyces* and *Schizosaccharomyces* here concerned, their calculated curves of growth are given in Figures 14 and 15.

In a mixed population of *Saccharomyces* and *Schizosaccharomyces* under anaerobic conditions the coefficients of the struggle for existence have the following values: α (showing the intensity of the influence of *Schizosaccharomyces* on *Saccharomyces*) $= 3.15$; β (intensity of the influence of *Saccharomyces* on *Schizosaccharomyces*) $= 0.439$. In other words, *one unit of volume of Schizosaccharomyces decreases the unutilized opportunity for growth of Saccharomyces 3.15 times as much as an equal unit of volume of Saccharomyces itself*. The species *Schizosaccharomyces* with its comparatively small volume takes up "a great number of places" in the microcosm. The reverse action of *Saccharomyces* on *Schizosaccharomyces* is comparatively weak. One unit of volume of *Saccharomyces* decreases the unutilized opportunity for growth of *Schizosaccharomyces* as much as 0.439 unit of the latter species' own volume.

(3) We now pass on to the most important part of this chapter, i.e., to the comparison of the empirically established coefficients of the struggle for existence with those which are to be expected on the basis of a direct study of the factors controlling growth. The values of the coefficients of the struggle for existence mentioned above are founded upon an analysis of the kinetics of growth of a mixed popula-

---

[4] In Chapter V we shall meet a more complicated situation.

tion. Let us at present leave them aside and endeavor to calculate the values of the coefficients of competition starting from the alcohol production. As mentioned above, the cessation of growth is connected with the reaching of a certain critical concentration of alcohol (characteristic for the given species under given conditions). Let us now assume that it is mainly alcohol that matters and that other by-products of fermentation are but of subordinate importance. Consequently, every unit of volume in each species produces a determined amount of alcohol, and when the latter reaches a certain threshold concentration the growth is checked. It follows that when a unit of volume of the first species produces an amount of alcohol considerably surpassing that produced by a unit of volume of the other species

TABLE V

Alcohol production in Saccharomyces cerevisiae and Schizosaccharomyces kephir
From Gause ('32b)

SACCHAROMYCES				SCHIZOSACCHAROMYCES			
Age in hours	Alcohol, per cent	Yeast volume in 10 c.c. of the medium	Alcohol per unit of yeast volume	Age in hours	Alcohol, per cent	Yeast volume in 10 c.c. of the medium	Alcohol per unit of yeast volume
16	1.100	10.20	0.108	48	0.728	3.08	0.236
16	0.480	5.33	0.090	72	1.425	5.51	0.259
24	1.690	12.22	0.138				
			Mean = 0.113				Mean = 0.247

$$\alpha_1 = \frac{0.247}{0.113} = 2.186$$

and the threshold values of alcohol in both are somewhat near to one another, the critical concentration of alcohol and the cessation of growth in the first species will be reached with a lower level of accumulated yeast volume. In Table V are given the data on the alcohol production in *Saccharomyces* and *Schizosaccharomyces* under anaerobic conditions. The determinations of the alcohol were made for the middle stages of growth, when its accumulation was almost strictly in proportion to the increase of the yeast volume. In *Saccharomyces* the alcohol production per unit of volume averages 0.113 per cent by weight, and in *Schizosaccharomyces* 0.247. These data show clearly that the latter species utilizes the medium unproductively and it occupies "a great number of places" by a comparatively small volume. At the same time this is an explanation of the low level of the accumulation of biomass in the separate cultures of *Schizosaccharomyces*, and the diminished volume of the mixed population in comparison with the volume of *Saccharomyces* cultivated separately.

We can now calculate approximately the critical concentrations of alcohol for the separate growth of each species of yeast if we multiply the maximal volumes of these species (K) by the alcohol production per unit of yeast volume. For *Saccharomyces* we shall have: $13.0 \times 0.113 = 1.47$, and for *Schizosaccharomyces*: $5.8 \times 0.247 = 1.43$. In other words, the critical alcohol concentrations for both species are about equal.

Let us now calculate the degree of influence of one species upon the unutilized opportunity for growth of another in a mixed population, or the coefficients of the struggle for existence. If we take as a unit the degree of decrease of the unutilized opportunity for growth of *Saccharomyces* by a unit of its own yeast volume, we have then to answer the following question: how much more or less does a unit of the yeast volume of *Schizosaccharomyces* decrease the unutilized opportunity for growth of *Saccharomyces* in the mixed population, in comparison with the effect of a unit of the volume of the latter species? Then, taking the ratio of the alcohol production per unit of yeast volume in *Schizosaccharomyces* to the alcohol production of *Saccharomyces* we shall find the coefficient of the struggle for existence according to the alcohol production: $\alpha = \dfrac{0.247}{0.113} = 2.186$. Correspondingly: $\beta = \dfrac{0.113}{0.247} = 0.457$.

(4) Comparing the results of the examination of the kinetics of growth of a mixed population with the data on the alcohol production, we observe a certain agreement in the general features. A very strong influence of *Schizosaccharomyces* upon *Saccharomyces* made apparent in the analysis of the kinetics of growth proved itself to be connected with the great alcohol production per unit of yeast volume in the former species. However, a strict coincidence of the data of these two independent methods of investigation does not occur here. Thus *Schizosaccharomyces* excretes a quantity of alcohol per unit of yeast volume 2.186 times as great as *Saccharomyces*, but influences the growth of the latter 3.15 times as much. Consequently, *Schizosaccharomyces* not only produces a greater amount of alcohol, but the

alcohol produced by it is so to say *"more toxic"* for *Saccharomyces* than the alcohol produced by the latter itself. All this tends to imply that the situation is here complicated by the influence of certain other waste products getting into the surrounding medium in small quantities. The relations between species in these experiments are therefore not so simple as has been supposed at the beginning of this section.

v

(1) The above described experiments of 1931 were repeated in 1932, and the new data confirmed all the observed regularities. In these new experiments the influence of oxygen upon the growth of a mixed population of the same two species of yeast was investigated, and this enabled us to further somewhat our understanding of the nature of the competitive process.

The experimental data given in the preceding section have to do with the growth of a yeast population under "anaerobic conditions," i.e., in test tubes. In order to study the influence of oxygen on the growth of the yeast population, together with experiments in test tubes we arranged other experiments under conditions of somewhat better aeration. The technique of such "aerobic" and "anaerobic" experiments has already been described at the beginning of this chapter. Here it must only be remarked that in the "aerobic" series the access of oxygen was very limited, and a part of the available energy was, as before, obtained by our species through alcoholic fermentation. As a result, a considerable amount of alcohol accumulated in the nutritive medium (as will be seen in the corresponding tables), and in its essential features the mechanism limiting the growth of the yeast population remained the same. The experiments of 1932 consisted of two aerobic and two anaerobic series. In them 168 separate microcosms were studied.

In all the experiments of 1932 nutritive medium of the same preparation was used. It was made according to the usual method, but the dry beer yeast was of another origin. As a result, the absolute values of growth were somewhat different. It must also be remarked that in all the new experiments the centrifuged volume of yeast was always reduced to 10 cm^3 of nutritive medium.

(2) Figure 16 represents the growth curves of *Saccharomyces*, *Schizosaccharomyces* and of the mixed population according to two

series of experiments in conditions analogous to the former anaerobic ones. The general character of these curves coincides with that of Figure 13. A more careful comparison of the anaerobic series of 1932 with that of 1931 shows that the first is characterized by considerably smaller absolute values of growth (Table VI). At the same time *Schizosaccharomyces* grown separately attains a somewhat higher level in comparison with *Saccharomyces* than formerly. Thus, the volume of the saturating population of the separately growing *Schizo-saccharomyces* represented in older experiments $\frac{5.8}{13.0} = 44.6$ per cent

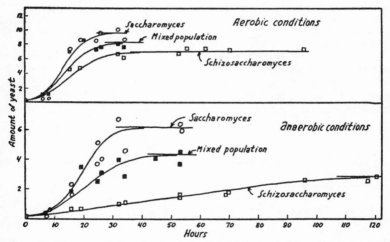

Fig. 16. The growth in volume of *Saccharomyces cerevisiae*, *Schizosaccharomyces kephir* and mixed population. Above: Aerobic conditions. Below: Anaerobic conditions (1932).

of that of *Saccharomyces* (1931), but in the new experiments it is $\frac{3.0}{6.25} = 48.0$ per cent (1932). In the experiments of 1932 the relative volume of *Schizosaccharomyces* in the mixed population increased also. As a result the decrease of the volume of the mixed population in comparison with the volume of separately growing *Saccharomyces* is more pronounced in 1932 than in 1931.

In spite of the alterations in the absolute values of growth and a certain change in the relative quantities of species, the coefficients of the struggle for existence which we had calculated for the anaerobic experiments of 1932 coincided almost completely with those of the year before. A similar coincidence exists in the ratio of the alcohol production of one species to that of another, which is to be found in Table VII. In this manner *the coefficients of the struggle for existence remain invariable under definite conditions in spite of the changing absolute values of growth.*

TABLE VI

Parameters of the logistic curves for separate growth of Saccharomyces cerevisiae and Schizosaccharomyces kephir under aerobic and anaerobic conditions (1932)

	K (MAXIMAL VOLUME)	b (COEFFICIENT OF GEOMETRIC INCREASE)	a (SEE APPENDIX II)	THE VALUE OF N AT $t = 0$
Saccharomyces anaerobic	6.25	0.21529	4.00652	0.112
Saccharomyces aerobic	9.80	0.28769	4.16358	0.152
Schizosaccharomyces anaerobic	3.0	0.04375	2.07234	0.335
Schizosaccharomyces aerobic	6.9	0.18939	2.78615	0.401

TABLE VII

Coefficients of the struggle for existence and the relative alcohol production under aerobic and anaerobic conditions

	COEFFICIENTS OF THE STRUGGLE FOR EXISTENCE		RELATIVE ALCOHOL PRODUCTION	
	α	β*	α_1	$\beta_1 = \dfrac{1}{\alpha_1}$
Anaerobic conditions (1931)	3.15	0.439	2.186	0.457
Anaerobic conditions (1932)	3.05	0.400	2.080	0.481
Aerobic conditions (1932)	1.25	0.850	1.25	0.80

* Here β does not coincide with $\dfrac{1}{\alpha}$.

(3) Let us now turn to the aerobic experiments (1932) and compare them to the anaerobic ones (1932). As might have been expected, in aerobic conditions the absolute values of growth of the yeast increase considerably (Fig. 16). What is especially striking is the behavior of *Schizosaccharomyces*. Though it is a slowly growing

species under anaerobic conditions, with a low level of biomass, it begins to grow rapidly with an access of oxygen and in its properties approaches *Saccharomyces*. The maximal volumes and coefficients of geometric increase given in Table VI show these regularities in a quantitative form. When there is no oxygen and fermentation is the only source of available energy, the coefficient of geometric increase in *Schizosaccharomyces* is very low and equal to 0.04375. Under the influence of oxygen this coefficient increases

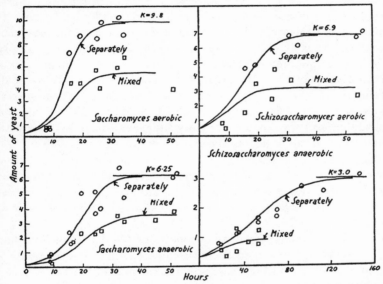

FIG. 17. The growth in volume of *Saccharomyces cerevisiae* and *Schizosaccharomyces kephir* cultivated separately and in the mixed population under aerobic and anaerobic conditions (1932). All curves are drawn according to equations.

4.3 times and attains 0.18939, whereas in *Saccharomyces* the coefficient of geometric increase under the same conditions rises but slightly (from 0.21529 to 0.28769).

The sharp changes in the properties of our species under aerobic conditions produce a completely new situation for the growth of a mixed population (see Fig. 17). As before, we have calculated the coefficients of the struggle for existence and Table VII shows that they differ considerably from the anaerobic ones. If in anaerobic experiments the coefficient α, which characterizes the intensity of influence of *Schizosaccharomyces* upon *Saccharomyces*, was equal to 3.05–3.15, than under aerobic conditions it is equal to 1.25. In other terms the influence of *Schizosaccharomyces* on *Saccharomyces* is no longer 3.05, but only 1.25 times as strong as the influence of the latter upon itself.

(4) Let us now examine the production of alcohol under aerobic conditions. The corresponding data are given in Table 2 (Appendix). As was to be expected, in aerobic conditions the amount of alcohol per unit of yeast volume is smaller than in anaerobic ones, because a part of the available energy is furnished by oxidation. It is interesting to compare the critical concentration of alcohol at which growth ceases, in aerobic and anaerobic conditions. Let us multiply as before the production of alcohol per unit of yeast volume by the maximal volume. For the anaerobic experiments of 1932 we shall obtain: *Saccharomyces*, $6.25 \times 0.245 = 1.53$; *Schizosaccharomyces*, $3.0 \times 0.510 = 1.53$. These threshold concentrations of alcohol coincide in both species, and they are sufficiently near to those with which we have had to deal in the anaerobic experiments of 1931. As to the threshold concentrations of alcohol in aerobic conditions, they prove to be higher than in the anaerobic ones, and in *Saccharomyces* the threshold lies somewhat higher than in *Schizosaccharomyces*: *Saccharomyces*, $9.80 \times 0.207 = 2.03$: *Schizosaccharomyces*, $6.9 \times 0.258 = 1.78$.

If we now calculate for aerobic conditions the degree of influence of *Schizosaccharomyces* upon *Saccharomyces* starting from the production of alcohol per unit of yeast volume, we shall obtain:

$$\alpha_1 = \frac{0.258}{0.207} = 1.25 \, .$$

Correspondingly the coefficient

$$\beta_1 = \frac{0.207}{0.258} = 0.80 \, .$$

Comparing these results with the data of the kinetics of growth, we see (Table VII) that in aerobic conditions the degree of influence of one species upon another calculated according to the system of equations of the struggle for existence fully coincides with the coefficients of the relative alcohol production. Therefore, the process of competition between our species in aerobic conditions is entirely regulated by alcohol, and there is scarcely any interference of other factors.

(5) We can now appreciate from a more general viewpoint the results of the aerobic experiments as well as those of this chapter. It has been shown that under aerobic conditions the theoretical equation of competition between two species of yeast for a common place in the microcosm given for the first time by Vito Volterra is completely realized. In other words, *if we know the properties of two species growing separately,* i.e., their coefficients of geometric increase, their maximal volumes, and alcohol production per unit of volume when alcohol limits the growth, *then connecting these values into a theoretical equation of the struggle for existence we can calculate in what proportion a certain limited amount of energy will be distributed between the populations of two competing species.* This means that we can calculate theoretically the growth of species and their maximal volumes in a mixed population. The equation of the struggle for existence expresses the idea that a potential geometric increase of each species in every infinitesimal interval of time is only realized up to a certain degree depending on the unutilized opportunity for growth at that moment, and that the species possesses certain coefficients of seizing this unutilized opportunity. Such theoretical calculations agree completely with the experimental data only under aerobic conditions, where the limitation of growth in both species depends almost completely on the ethyl alcohol. In the case of anaerobic conditions the situation becomes more complicated as a result of the influence of certain other waste products. This shows that extreme care is necessary in the investigation of biological systems, because various and often unexpected factors may participate in the process of interaction between two species.

CHAPTER V

COMPETITION FOR COMMON FOOD IN PROTOZOA

I

(1) At the end of the last century Boltzmann, considering the struggle for existence in the biosphere as a whole, remarked that there exists a considerable quantity of essential mineral substances needed by all living beings, but that the resources of available solar energy are comparatively more restricted and they constitute the narrow link representing the principal object of competition. This circumstance has since been pointed out by many biophysicists, and we will quote the words of Boltzmann himself ('05): "The general struggle for existence of all living beings is not the struggle for the fundamental substances, for these fundamental substances indispensable for all living creatures exist abundantly in the air, the water and the soil. This struggle is not a struggle for the energy which in the form of heat, unfortunately not utilizable, is present in a great quantity in every object, but it is a struggle for entropy, which is available when energy passes from the hot sun to the cold earth. In order to utilize in the best manner this passage, the plants spread under the rays of the sun the immense surface of their leaves, and cause the solar energy before reaching the temperature level of the earth to make syntheses of which as yet we have no idea in our laboratories. The products of this chemical kitchen are the object of the struggle in the animal world." This idea of Boltzmann that the available solar energy represents the narrow link for the living matter in the biosphere taken as a whole is in a certain agreement with the data of the modern geochemists. Thus Professor Vernadsky ('26) points out that a part of the solar energy which is capable of producing chemical work on the earth is to the very end utilized in the mechanism of the biosphere. In other words, the transforming surface of the green living matter utilizes entirely the rays of a definite wave-length in the process of photosynthesis.

(2) However that may be, the energetic side of the struggle for existence in the biosphere as a whole has as yet been little studied,

and at the present level of our knowledge we will have to undertake a detailed analysis of the most simple cases only. But the words of Boltzmann compel us to turn our particular attention to those narrow links in the conditions of our microcosm which constitute the real objects of the competition. In the foregoing chapter we had to deal with a competition of the yeast cells for the utilization of a certain limited amount of energy in the test tube. The limit for the growth of the biomass in these organisms was connected with the accumulation of the waste product (alcohol), and this factor stopped the growth before the exhaustion of the energetic resources of the microcosm. As a result the entire process of competition could be expressed in terms of the narrow link—alcohol production. Thus the situation in these experiments was a peculiar one.

In the present chapter we will try to approach the regularities which, as Boltzmann supposed, are characteristic for the biosphere as a whole. We will examine the struggle for existence in carefully controlled populations of Protozoa. Here the growth will be limited by an insufficiency of organic nutritive substances, a factor analogous to an insufficiency of available energy. The second peculiarity is that the energetical resources of the microcosm will be maintained continuously at a certain fixed level in the course of the experiment. This approaches somewhat to what exists in nature, where the level of energy is maintained by the uninterrupted influx of solar energy. As before we will be concerned in these experiments with the problem in what proportion the energy of the microcosm will be distributed between the populations of the two competing species. But besides this first stage we shall be enabled to examine here the following fundamental question: *Will one species drive out the other after all the available energy of the microcosm has been already taken hold of?* And if so, *will one species in these conditions drive the other one out completely,* or will a certain equilibrium become established between them?

(3) It has been already tried more than once to use Protozoa for the study of the communities of organisms and their succession under laboratory conditions. But as an ecologist has recently remarked, the mere fact of a community set up in a laboratory dish does not mean at all that it is simple. Interesting observations have been made on the succession of communities of Protozoa in a hay infusion by Woodruff ('12) Skadowsky ('15) and more recently by Eddy ('28).

However, in experiments of this type there exists a great number of different factors not exactly controlled, and a considerable difficulty for the study of the struggle for existence is presented by the continuous and regular changes in the environment. It is often mentioned that one species usually prepares the way for the coming of another species. Recollecting what we have said in Chapter II it is easy to see that in such a complicated environment it is quite impossible to decide how far the supplanting of one species by another depends on the varying conditions of the microcosm which oppress the first species, and in what degree this is due to direct competition between them. In this connection one of the main problems of our experiments with Protozoa has been to eliminate the complicating influence of numerous secondary factors, and to apply such a technique of cultivation as would enable one to form a perfectly clear idea as to the nature of the factor limiting growth. This could not be done at once and the technique of our first experiments presented all the usual defects. Only later, taking into account certain suggestions of American authors, we made use of a synthetic medium for cultivating the Protozoa, and the result furnished exceedingly clear data to a detailed description of which we will soon pass.

(4) A new property of the infusorian population distinguishing it from that of yeast cells is that the infusorian population constitutes a secondary population living at the expense of bacteria which it devours. Thus here appears an elementary food chain: bacteria → infusoria. In our initial experiments the standardization of the conditions of cultivation was only a quite superficial one. Without taking any precautions as to an exact control of the physicochemical properties of the medium and the number of growing bacteria, we prepared the nutritive medium in the following manner: to 100 c.c. of tap water 0.5 gr. of oatmeal was added; the whole was boiled for 10 minutes, left to stand and then the upper liquid was carefully poured off, diluted 1½ times by water, and sterilized in an autoclave. After this an inoculation of *Bacillus subtilis* was made, and the medium was put into the thermostat at 32° for seven days in order to obtain an abundant growth of bacteria. Before using, the medium was diluted twice by tap-water, and without any further sterilization was put into test tubes. (This was the so-called "oaten medium without sediment." The "oaten medium with sediment" mentioned in Chapter VI differs in its not being diluted by water before using, and

a small quantity of sediment originating from the oatmeal was allowed to remain.) The cultivation was made in tubes with a flat bottom (about 1 cm. in diameter and 5–6 cm. high) of the nutritive solution. The tubes were closed by cotton wool stoppers and kept in a moist thermostat at 26°C. Close paraffinized cork stoppers were not found convenient because if we use them the population begins to die off immediately after cessation of growth, and the curves take the form described by Myers ('27). At the same time under optimal conditions after the growth of the population has ceased the level of the population is maintained unchanged for a certain time, and only later Paramecia begin to die off.

In the initial experiments no change of the medium in the course of growth of the population was made, and the increase in the number of individuals was studied according to the average values for the test tubes of a definite age. The contents of the tube was destroyed every time after examination just as in the experiments with yeast. The counting was made under a magnifying glass on a slide plate. Figure 18 represents the growth of the number of individuals in pure lines of *Paramecium caudatum* and *Stylonychia mytilus* cultivated separately and in a mixed population. These data are founded on two experiments which gave similar results. At the beginning of the experiment into each tube were placed five *Paramecium*, or five *Stylonychia*, or five *Paramecium* plus five *Stylonychia* in the case of a mixed population. *Stylonychia* for inoculation must be taken from young cultures to avoid an inoculation of degenerating individuals.

(5) The growth curves of the number of individuals in Figure 18 are S-shaped and resemble our well known yeast curves. After growth has ceased the level of the saturating population is maintained for a short time, and then begins the dying off of the population which is particularly distinct in *Stylonychia*. It is evident that this dying off is regulated by factors quite different from those which regulate growth, and that a new system of relations comes into play here. Therefore there is no reason to look for rational equations expressing both the growth and dying off of the populations.

Figure 18 shows that *Stylonychia*, and especially *Paramecium*, in a mixed culture attain lower levels than separately. The calculated coefficients of the struggle for existence have the following values: α (influence of *Stylonychia* on *Paramecium*) = 5.5 and β (influence of *Paramecium* on *Stylonychia*) = 0.12. This means that *Stylonychia* influences *Paramecium* very strongly, and that every individual of the former occupies a place available for 5.5 Paramecia. With our technique of cultivation it is difficult to decide on what causes this depends. As a supposition only one can point to food consumption.

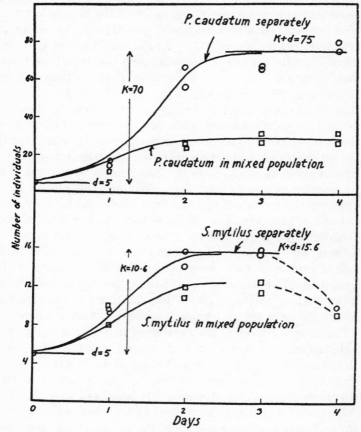

Fig. 18. The growth in number of individuals of *Paramecium caudatum* and *Stylonychia mytilus* cultivated separately and in the mixed population. *d* denotes lower asymptote. From Gause ('34b).

(6) We have but to change slightly the conditions of cultivation and we shall obtain entirely different results. Figure 19 represents the growth of populations of the same species on a dense "oaten medium with sediment" sown with various wild bacteria. Here

owing to an increase in the density of food the absolute values of the maximal population in both species have considerably increased. The character of growth of the mixed population now essentially differs from the former one: *Paramecium* strongly influences *Stylonychia*, while *Stylonychia* has almost no influence upon *Paramecium*. We simply have here an "alchemical stage" of investigation, and the absence of an exact control of the conditions of the medium creates the impression of a complete arbitrariness of the results of our experiments. Unfortunately many protozoological researches are still in this stage, and the idea is very widespread that "Protozoa are not

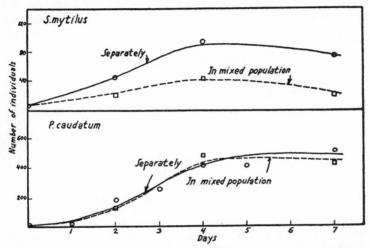

FIG. 19. The growth in number of individuals of *Paramecium caudatum* and *Stylonychia mytilus* cultivated separately and in the mixed population. Medium contains wild bacteria.

entirely satisfactory for the study of populations as they require bacteria for food, and it is very difficult to measure accurately and to analyze the relations between protozoan population and the bacterial population."

In order to draw any reliable conclusions as to the quantitative laws of the struggle for existence in Protozoa we must begin by elaborating the technique of cultivation, keeping in mind the following excellent words of Raymond Pearl: "When the biologist exercises something approaching the same precision and infinitely painstaking care, over *all* the most trivial details of a biological experiment that the physicist does over his, the results tend to take on a degree of precision and uniformity not so far short of that usual in the older science, as we are accustomed to expect" ('28, p. 35).

II

(1) Jennings pointed out the necessity of a careful control of bacteria in the cultures of Protozoa in 1908, and one of the first attempts to grow Paramecia in pure cultures of bacteria was made by Hargitt and Fray ('17), Oehler ('20) and Jollos ('21). Since then numerous researches have appeared and a good review of them can be found in the recently published book of Sandon ('32) *The Food of Protozoa* as well as in Hartmann ('27) and Belar ('28). It can be noted in a quite general form that in order to standardize the conditions of cultivation

TABLE VIII
Balanced physiological salt solution of Osterhout

NaCl..	2.35 gr.
$MgCl_2$..	0.184 gr.
$MgSO_4$..	0.089 gr.
KCl..	0.050 gr.
$CaCl_2$..	0.027 gr.
Bidistilled water to 100 c.c.	

This solution is diluted with bidistilled water 225 times.

of Protozoa it is necessary: (1) to standardize the quality of food—to cultivate the Protozoa on bacteria of a definite species, (2) to standardize the quantity of food—the number of bacteria per unit of volume must have a fixed value, and (3) to standardize the physicochemical conditions of the medium. These difficult and as one may think hardly realizable problems have been solved very simply for *Oxytricha* by Johnson ('33), who has been partly preceded by Barker and Taylor ('31). The method is this: a culture of a certain bacterium is made on a solid medium and then a fixed quantity of bacteria is taken off the solid medium and transferred into a balanced physiological salt solution, where these bacteria do not multiply and serve as food for the Protozoa. Like Johnson we used Osterhout's salt solution, the composition of which is given in Table VIII. As will be shown further on, in certain experiments this medium was buffered and kept at a definite hydrogen ion concentration (pH). Special experiments made by Johnson showed that the bacteria do not multiply in this medium, and that their number scarcely changes within 24 hours.

Beginning our experiments on the growth of pure and mixed populations of *Paramecium caudatum*, *Paramecium aurelia* and *Stylonychia pustulata*, we devoted a certain time to finding a culture of bacteria suitable as food for all three species of Protozoa. Using the data published by Philpot ('28) we chose finally the pathogenic bacterium *Bacillus pyocyaneus*, which was cultivated in Petri dishes at 37°C. on a solid medium of the following composition: peptone, 1 gr.; glucose, 2 gr.; K_2HPO_4, 0.02 gr.; agar-agar, 2 gr., per 100 cm^3 of tap-water. One standardized uniformly filled platinum loop of fresh *Bacillus pyocyaneus* taken off the solid medium was placed in 10 cm^3 of Osterhout's salt solution. This mixture was prepared anew every day, and we will speak of it as the "one-loop" medium.

(2) Such a standard and convenient technique of cultivation enables us to approach the experimental investigation of an important problem: the course of the process of competition for a source of energy kept continually at a certain level. With this object the cultivation was carried on in graduate tubes for centrifugation of 10 c.c. capacity, which were filled with nutritive medium up to 5 c.c. and closed with cotton wool stoppers. Twenty individuals of the corresponding species were placed in every tube, or 20 plus 20 in case of a mixed culture. The medium was changed daily in the following manner. The tube was placed in a centrifuge, and after two minutes of centrifugation with 3500 revolutions per minute the infusoria fell to the bottom, the liquid above was very gently drawn off by means of a pipette with a caoutchouc ball and a freshly made nutritive medium was poured in. Besides this, every other day each culture was washed with the salt solution free of bacteria, in order to prevent the accumulation of waste products in the few drops of liquid remaining at the bottom of the tube with the Paramecia at the moment when the medium was changed. For this purpose after the pouring off of the old medium the tubes were filled with a pure salt solution, centrifuged and the liquid was drawn off a second time. Every day before the medium was changed each culture was carefully stirred up, 0.5 c.c. of the liquid was taken out and the number of infusoria in it counted. After counting the sample was destroyed. All

the experiments were made in a moist thermostat at 26°C. with pure lines of infusoria.

(3) In an experiment of such a type all the properties of the medium are brought to a certain invariable "standard state" at the end of every 24 hours. Hence, we acquire the possibility of investigating the following problem: can two species exist together for a long time in such a microcosm, or will one species be displaced by the other entirely? This question has already been investigated theoretically by Haldane ('24), Volterra ('26) and Lotka ('32b). It appears that the properties of the corresponding equation of the struggle for existence are such that if one species has any advantage over the other it will inevitably drive it out completely (Chapter III). It must be noted here that it is very difficult to verify these conclusions under natural conditions. For example, in the case of competition between

TABLE IX

Contents of the microcosms in the experiments with Osterhout's medium

CONTENTS OF THE MICROCOSM	NUMBER OF MICROCOSMS
(1) *Paramecium caudatum* separately......................	4
(2) *Stylonychia pustulata* separately.......................	5
(3) *Paramecium aurelia* separately.........................	3
(4) *P. caudatum* + *P. aurelia*.............................	3
(5) *P. caudatum* + *S. pustulata*...........................	3
(6) *P. aurelia* + *S. pustulata*.............................	3

two species of crayfish (Chapter II) a complete supplanting of one species by another actually takes place. However, there is in nature a great diversity of "niches" with different conditions, and in one niche the first competitor possessing advantages over the second will displace him, but in another niche with different conditions the advantages will belong to the second species which will completely displace the first. Therefore side by side in one community, but occupying somewhat different niches, two or more nearly related species (e.g., the community of terns, Chapter II) will continue to live in a certain state of equilibrium. There being but a single niche in the conditions of the experiment it is very easy to investigate the course of the displacement of one species by another.

(4) Two series of experiments were arranged by us in which the

process of competition was studied in 21 microcosms for a period of 25 days. Table IX shows the combinations of separate species of Protozoa which were used. Let us first of all analyze the competition between *Paramecium caudatum* and *Paramecium aurelia*. The data on the growth of pure and mixed populations of these species are presented in Table 3 (Appendix) which gives the number of individuals in a sample of 0.5 cm³ taken from a culture of 5 cm³ in volume. (A separate counting of the number of individuals in every culture was discontinued from the twentieth day, and we began to take average samples from the similar cultures.)

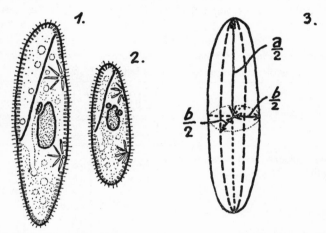

Fig. 20. *Paramecium caudatum* (1) and *Paramecium aurelia* (2) according to Kalmus ('31). (3) Measurements for the calculation of volume of *Paramecium*.

In order to investigate the process of competition, we had to pass from the number of individuals of *P. caudatum* and *P. aurelia* to their biomasses, as these species differ rather strongly in size (see Fig. 20). In order to obtain an idea of the biomass we had recourse to the volumes of these species. *P. caudatum* and *P. aurelia* were measured under the conditions of our experiments (Table X) and on the basis of these measurements the volumes were calculated. As in shape *Paramecium* after fixation approaches somewhat closely to an ellipsoid of rotation with the half-axes: $\frac{a}{2}, \frac{b}{2}, \frac{b}{2}$ (see Fig. 20), the calculation of the volumes was made according to the formula for this body.

Taking the volume of the larger *P. caudatum* equal to unity, the volume of *P. aurelia* can be easily expressed in a relative form.   Under different conditions the relative volume of *P. aurelia* varies somewhat, but for Osterhout's medium it can be taken as equal on an average to 0.39 of the volume of *P. caudatum*.   In this way, in order to pass from the growth of the number of individuals of the two species of Paramecia to the growth of their volumes, we can leave without alteration the number of individuals of *P. caudatum*, and only *diminish the number of individuals* of the small *P. aurelia* by multiplying it in every case by 0.39.

(5) Figure 21 represents graphically the growth in the number of

TABLE X

Measurements of Paramecium caudatum and Paramecium aurelia (after fixation)
Specification of measurements is taken from Figure 20

ORIGIN OF PARAMECIA	AVERAGE VALUES FOR P. CAUDATUM IN DIVISIONS OF OCULAR-MICROMETER	AVERAGE VALUES FOR P. AURELIA IN DIVISIONS OF OCULAR-MICROMETER	CALCULATED VOLUME OF P. AURELIA (VOLUME OF P. CAUDATUM = 1)
Growing culture with Osterhout's medium..........................	$a = 18.3$ $b = 6.0$	$a = 13.8$ $b = 4.2$	$0.39 \begin{cases} 0.37 \\ 0.41 \end{cases}$
Old culture with Osterhout's medium..	$a = 17.1$ $b = 5.1$	$a = 12.6$ $b = 3.8$	
Culture with the buffered medium....	$a = 18.0$ $b = 6.2$	$a = 12.9$ $b = 4.8$	0.429

individuals and in the volumes of *P. caudatum* and *P. aurelia* cultivated separately in a medium changed daily for 25 days.   The general character of the curves shows that the growth of population under these conditions has an S-shaped form.   At a certain moment the possibility of growth in a given microcosm is apparently exhausted, and with a continuously maintained level of nutritive resources a certain equilibrium of population is established.   The oscillations of population round this state of equilibrium are not governed by any apparent law, and depend on various accidental causes (variation in temperature of the thermostat, a slight variability in the composition of the synthetic medium, etc.).   A comparison of the curves of growth of *P. caudatum* and *P. aurelia* shows that as regards the number of individuals the level of the saturating population of *P. aurelia* is considerably higher than that of *P. caudatum*. Nevertheless, the comparison of the volumes shows something completely different; in this respect *P. aurelia* only slightly surpasses *P. caudatum*, accumulating at the expense of a certain definite level of food resources a scarcely larger biomass. As will be shown further on, the Osterhout salts medium is not quite favorable in its properties

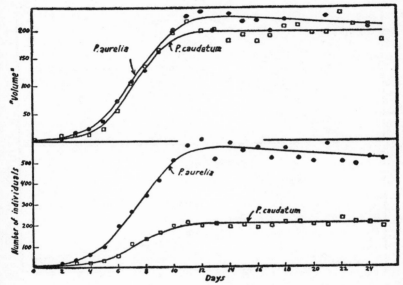

FIG. 21. The growth of the number of individuals and of the "volume" in *Paramecium caudatum* and *Paramecium aurelia* cultivated separately on the medium of Osterhout. From Gause ('34d).

for the Paramecia, and this complicates the question as to the factors limiting growth. On the one hand, the insufficiency of food plays a part here which we can judge of by a direct observation of the cultures: with a population in equilibrium the turbid bacterial medium introduced daily becomes quite transparent after a certain time, as the bacteria are entirely devoured by the Paramecia. However, owing to a comparatively high concentration of bacteria and a somewhat unoptimal reaction of the medium a depressory action of certain other influences plays also a rôle here.

(6) The data on the growth of the volumes of *P. caudatum* and *P. aurelia* in a mixed population are given in Figure 22. The curves of growth of each species in a mixed culture are presented here on the background of control curves corresponding to the free growth of the same species. It is easy to see that the growth of a mixed population consists of two periods: (a) during the first period (till the eighth day), the species grow and compete for the seizing of the still unutilized energy (food resources). But the moment approaches gradually

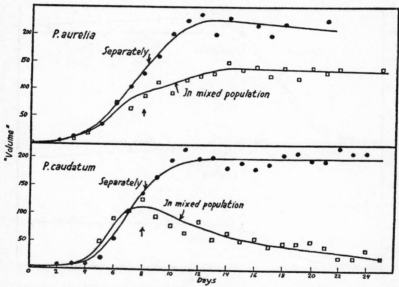

FIG. 22. The growth of the "volume" in *Paramecium caudatum* and *Paramecium aurelia* cultivated separately and in the mixed population on the medium of Osterhout. From Gause ('34d).

when all the utilizable energy is already taken hold of, and the total of the biomasses of the two species tends to reach the maximal possible biomass under given conditions. (This happens on the eighth day; the total biomass is equal to about 210.) This first period corresponds to what we have already observed in yeast cells. (b) After this there can only arise *the redistribution of the already seized energy between the two species*, i.e. the displacement of one species by another. Figure 22 shows that such a displacement is actually observed in the experiment: the number of *P. caudatum* gradually diminishes as a result of its being driven out by *P. aurelia.* As several further experiments have shown (see Fig. 24), the process of competition under our conditions has always resulted in one species being entirely displaced by another, in complete agreement with the predictions of the mathematical theory.

If we consider the curves in Figure 22 more in detail, we shall note that they generally are of a rather complicated character. It is interesting to note that *P. caudatum* in a mixed culture at the beginning of the experiment grows even better than separately. This is apparently a consequence of the more nearly optimal relationships between the density of the Paramecia and that of the bacterial food, in accordance with the observations of Johnson ('3ᴐ).

III

(1) Although the situation in our experiments with Osterhout's medium has been considerably simpler than in the case of the "oaten medium," it is still too complicated for a clear understanding of the mechanism of competition. In fact, why has one species been victorious over another? In the case of yeast cells we answered that the success of the species during the first stage of competition depends on definite relations between the coefficients of multiplication and the alcohol production, and that it can be exactly predicted with the aid of an equation of the struggle for existence. What will be our answer for the population of Paramecia?

To investigate this problem we made the conditions of the experimentation the next step in the simplification. We endeavored to make a medium with a very small concentration of nutritive bacteria and optimal in its physicochemical properties for Paramecia. Under such conditions the competition for common food between two species of Protozoa has been reduced to its simplest form.

(2) As Woodruff has shown ('11, '14), the waste products of Paramecia can depress the multiplication and be specific for a given species. In any case we are very far from an exact knowledge of their rôle and chemical composition. Therefore first of all we must eliminate the complicating influence of these substances. This problem is the reverse of the one we had to do with in the preceding chapter. There in the experiments with yeast we tried to set up conditions under which the food resources of the medium should be very considerable at the time when the concentration of the waste

products had already attained a critical value. Now with Paramecia our object is that the concentration of the waste products should still be very far from the critical threshold at the moment when the food is exhausted.

First of all we turned our attention to the hydrogen ion concentration (pH), which in the light of the researches of Darby ('29) can be of great importance for our species. When Paramecia are cultivated in Osterhout's medium, pH is near to 6.8 and unstable, where-

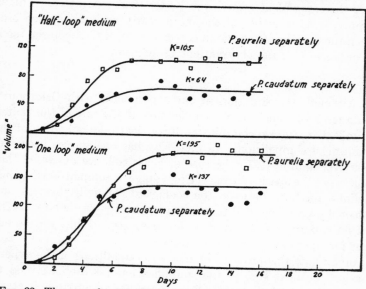

Fig. 23. The growth of the "volume" in *Paramecium caudatum* and *Paramecium aurelia* cultivated separately on the buffered medium ("half-loop" and "one-loop" concentrations of bacteria). From Gause ('34d).

as the reaction in our wild cultures is commonly near to 8.0. Therefore we, like Johnson, buffered Osterhout's medium by adding 1 cm³ of $\frac{m}{20}$ KH_2PO_4 to 30 cm³ of diluted salt solution, and bringing the reaction of the medium with the aid of $\frac{m}{20}$ KOH to pH = 8.0. At the same time we isolated new pure lines of Paramecia out of our wild culture, as the Paramecia which had been cultivated for a long time on Osterhout's medium could not stand a sudden transfer into a buffered medium.

In order to diminish the concentration of the bacteria we made a new smaller standard loop for preparing the "one-loop medium," and also arranged experiments in which the one loop medium was diluted twice ("half-loop medium").   The data obtained are given in Table 4 (Appendix) where every figure represents a mean value from the observations of two microcosms.   This material is represented graphically in Figures 23, 24 and 25.

Let us examine Figure 23.   The curves of growth of pure populations of *P. caudatum* and *P. aurelia* with different concentrations of the bacterial food show that the lack of food is actually a factor limiting growth in these experiments.   With the double concentration of food the volumes of the populations of the separately growing species also increase about twice (from 64 up to 137 in *P. caudatum*; 64 × 2 = 128; from 105 up to 195 in *P. aurelia*; 105 × 2 = 210).   Under these

TABLE XI

Parameters of the logistic curves for separate growth of Paramecium caudatum and Paramecium aurelia

Buffered medium with the "half-loop" concentration of bacteria

	P. AURELIA	P. CAUDATUM
Maximal volume (K)	$K_1 = 105$	$K_2 = 64$
Coefficient of geometric increase (b)	$b_1 = 1.1244$	$b_2 = 0.7944$

conditions the differences in the growth of populations of *P. aurelia* and *P. caudatum* are quite distinctly pronounced: the growth of the biomass of the former species proceeds with *greater rapidity*, and it accumulates a *greater biomass than P. caudatum at the expense of the same level of food resources*.[1]   If we now express the curves of separate growth of both species under a half-loop concentration of bacteria with the aid of logistic equations we shall obtain the data presented in Table XI.   This table shows clearly that *P. aurelia* has perfectly definite advantages over *P. caudatum* in respect to the basic characteristics of growth.

(3) We will now pass on to the growth of a mixed population of *P. caudatum* and *P. aurelia*.   The general character of the curves on

[1] This is apparently connected with the resistance of *P. aurelia* to the waste products of the pathogenic bacterium, *Bacillus pyocyaneus* (see Gause, Nastukova and Alpatov, '35).

Figures 22, 24 and 25 is almost the same, but there are certain differences concerning secondary peculiarities. For a detailed acquaintance with the properties of a mixed population we will consider the growth with a half-loop concentration of bacteria (Fig. 24). First of all we see that as in the case examined before the competition between our species can be divided into two separate stages: up to the fifth

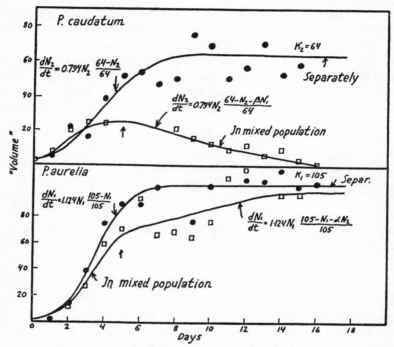

Fig. 24. The growth of the "volume" in *Paramecium caudatum* and *Paramecium aurelia* cultivated separately and in the mixed population on the buffered medium with the "half-loop" concentration of bacteria. From Gause ('34d).

day there is a competition between the species for seizing the so far unutilized food energy; then after the fifth day of growth begins the redistribution of the completely seized resources of energy between the two components, which leads to a complete displacement of one of them by another. The following simple calculations can convince one that on the fifth day all the energy is already seized upon. At the expense of a certain level of food resources which is a constant one in all "half-loop" experiments and may be taken as unity, *P. aurelia* growing separately produces a biomass equal to 105 volume units, and *P. caudatum* 64 such units. Therefore, one unit of volume of *P. caudatum* consumes $\frac{1}{64} = 0.01562$ of food, and one unit of volume

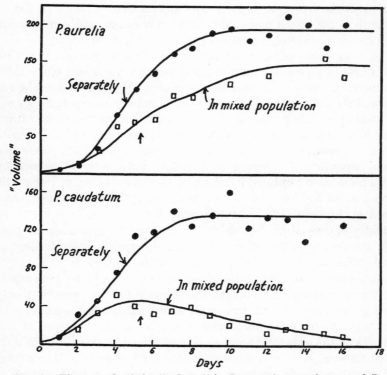

FIG. 25. The growth of the "volume" in *Paramecium caudatum* and *Paramecium aurelia* cultivated separately and in the mixed population on the buffered medium with the "one-loop" concentration of bacteria.   From Gause ('34d).

of *P. aurelia* $\frac{1}{105} = 0.00952$.   In other words, one unit of volume of *P. caudatum* consumes 1.64 times as much food as *P. aurelia*, and the food consumption of one unit of volume in the latter species constitutes but 0.61 of that of *P. caudatum*. These coefficients enable us to recalculate the volume of one species into an equivalent in respect to the food consumption volume of another species.

On the fifth day of growth of a mixed population the biomass of
P. caudatum (in volume units) is equal to about 25, and of *P. aurelia*
to about 65.   If we calculate the total of these biomasses in equivalents of *P. aurelia*, we shall have: $(25 \times 1.64) + 65 = 106$ (maximal
free growth of *P. aurelia* is equal to 105).   The total of the biomasses
expressed in equivalents of *P. caudatum* will be $(65 \times 0.61) + 25 =$
65 (with the free growth 64).   This means that on the fifth day of
growth of the mixed population the food resources of the microcosm
are indeed completely taken hold of.

(4) The first period of competition up to the fifth day is not all
so simple as we considered it in the theoretical discussion of the third
chapter, or when examining the population of yeast cells.   The nature of the influence of one species on the growth of another does not
remain invariable in the course of the entire first stage of competition,
and in its turn may be divided into two periods.   At the very beginning *P. caudatum* grows even somewhat better in a mixed population
than separately (analogous to Fig. 22), apparently in connection
with more nearly optimal relations between the density of Paramecia
and that of the bacteria in accordance with the already mentioned
data of Johnson ('33).   At the same time *P. aurelia* is but very
slightly oppressed by *P. caudatum*.   As the food resources are used
up, the Johnson effect disappears, and the species begin to depress
each other as a result of competition for common food.

It is easy to see that all this does not alter in the least the essence
of the mathematical theory of the struggle for existence, but only
introduces into it a certain natural complication: the coefficients of
the struggle for existence, which characterize the influence of one
species on the growth of another, do not remain constant but in their
turn undergo regular alterations as the culture grows.   The curves
of growth of every species in a mixed population in Figure 24 up to
the fifth day of growth have been calculated according to the system
of differential equations of competition with such varying coefficients.
In the first days of growth the coefficient β is negative and near to
-1, i.e., instead of $-\beta N_1$ we obtain $+N_1$.   In other words, the
presence of *P. aurelia* does not diminish, but increases the possibility of growth of *P. caudatum*, which proceeds for a certain time
with a potential geometrical rate, outrunning the control culture
$\left(\dfrac{64 - N_2 + N_1}{64} \text{ remains near to unity} \right)$.   At this time the coefficient

α is equal to about $+0.5$; in other words, *P. aurelia* suffers from a slight depressing influence of *P. caudatum*. Later the inhibitory action of one species upon the growth of another begins to manifest itself more and more in proportion to the quantity of food consumed, because the larger is the part of the food resources already consumed the less is the unutilized opportunity for growth. In our calculations for *P. caudatum* from the second and for *P. aurelia* from the fourth days of growth we have identified the coefficients of competition with the coefficients of the relative food consumption, i.e., $\alpha = 1.64$, $\beta = 0.61$. It is obvious that this is but a first approximation to the actual state of things where the coefficients gradually pass from one value to another. The entire problem of the changes in the coefficients of the struggle for existence in the course of the growth of a mixed population (which apparently are in a great measure connected with the fact that the Paramecia feed upon living bacteria) needs further detailed investigations on more extensive experimental material than we possess at present.

(5) It remains to examine the second stage of the competition, i.e., the direct displacement of one species by another. An analysis of this phenomenon can no longer be reduced to the examination of the coefficients of multiplication and of the coefficients of the struggle for existence, and we have to do in the process of displacement with a quite new qualitative factor: the rate of the stream which is represented by population having completely seized the food resources. As we have already mentioned in Chapter III, after the cessation of growth a population does not remain motionless and in every unit of time a definite number of newly formed individuals fills the place of those which have disappeared during the same time. Among different animals this can take place in various ways, and a careful biological analysis of every separate case is here absolutely necessary. In our experiments the principal factor regulating the rapidity of this movement of the population that had ceased growing was the following technical measure: a sample equal to $\frac{1}{10}$ of the population was taken every day and then destroyed. In this way a regular decrease in the density of the population was produced and followed by the subsequent growth up to the saturating level to fill in the loss.

During these elementary movements of thinning the population and filling the loss, the displacement of one species by another took place. The biomass of every species was decreased by $\frac{1}{10}$ daily.

Were the species similar in their properties, each one of them would again increase by $\frac{1}{10}$, and there would not be any alteration in the relative quantities of the two species. However, as one species grows quicker than another, it succeeds not only in regaining what it has lost but also in seizing part of the food resources of the other species. Therefore, every elementary movement of the population leads to a diminution in the biomass of the slowly growing species, and produces its entire disappearance after a certain time.

(6) The recovery of the population loss in every elementary movement is subordinate to a system of the differential equations of competition. In the present stage of our researches we can make use of these equations for only a qualitative analysis of the process of displacement. They will show us exactly what particular species in the population will be displaced. However, the quantitative side of the problem, i.e., the rate of the displacement, still requires further experimental and mathematical researches and we will not consider it at present.

The qualitative analysis consists in the following. Let us assume that the biomass of each component of the saturating population is decreased by $\frac{1}{10}$. Then according to the system of differential equations, inserting the values of the coefficients of multiplication and of the coefficients of food consumption, we shall be able to say how each one of the components can utilize the now created possibility for growth. The result of the calculations shows that *P. aurelia*, primarily owing to its high coefficient of multiplication, has an advantage and increases every time comparatively more than *P. caudatum*.[2]

In summing up we can say that in spite of the complexity of the process of competition between two species of infusoria, and as one may think a complete change of conditions in passing from one period of growth to another, a certain law of the struggle for existence which may be expressed by a system of differential equations of competition remains invariable all the time. The law is that the species possess definite potential coefficients of multiplication, which are realized at

---

[2] It is obvious that in these calculations it is necessary to introduce varying coefficients of the struggle for existence. At the same time with our technique of cultivation corrections to the "elementary movements" must be also included in an analysis of the first stage of growth of a mixed population (an approximation to the asymptote). But at the present stage of our researches we have neglected them.

every moment of time according to the unutilized opportunity for growth. We have only had to change the interpretation of this unutilized opportunity.

(7) It seems reasonable at this point to coördinate our data with the ideas of the modern theory of natural selection. It is recognized that fluctuations in numbers resembling the dilutions we have artificially produced in our microcosms play in general a decisive rôle in the removal of the less fitted species and mutations (Ford, '30). An interesting mathematical expression of this process proposed by Haldane ('24, '32) can be formulated thus: how does the rate of increase of the favorable type in the population depend on the value of the coefficient of selection k? In its turn the coefficient of selection characterizes an elementary displacement in the relation between the two types per unit of time—one generation. Therefore the problem resolves itself into a determination of the functional relationship between the increase of concentration of the favorable type and the elementary displacement in its concentration. A recent theoretical paper by Ludwig ('33) clearly shows how the fluctuation in the population density alters the relation between the two types owing to the fact that one of them has a somewhat higher probability of multiplication than the other. It seems to us that there is a great future for the Volterra method here, because it enables us not to begin the theory by the coefficient of selection but to calculate theoretically the coefficient itself starting from the process of interaction between the two species or mutations.

IV

(1) How complicated are processes of competition under the conditions approaching those of nature can be seen from the experiments made by Gause, Nastukova and Alpatov ('35). They studied the influence of biologically conditioned media on the growth of a mixed population of *Paramecium caudatum* and *P. aurelia*. The analysis of the relative adaptation of the two species at different stages of population growth has shown that *P. caudatum* has an advantage over *P. aurelia* in the coefficient of geometric increase (in the absence of *Bacillus pyocyaneus* in the nutritive medium which in the experiments described above inhibited *P. caudatum* by its waste products) whilst *P. aurelia* surpasses *P. caudatum* in the resistance to waste products. Therefore if the decisive factor of competition is a rapid

utilization of the food resources, *P. caudatum* has an advantage over *P. aurelia*; but if the resistance to waste products is the essential point, then *P. aurelia* will take place of *P. caudatum*.

It is interesting to note also that in the complicated situation of these experiments the superiority of one species over another in competition did not simply reflect the properties of these species taken independently, but *was often essentially modified by the process of their interaction.*

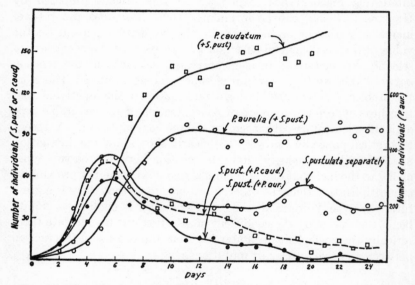

Fig. 26. The growth of the number of individuals of *Stylonychia pustulata* cultivated separately, and in the mixed populations with *Paramecium caudatum* and *Paramecium aurelia* (on the medium of Osterhout).

(2) If we turn to the population growth of *Stylonychia pustulata* and its competition with two species of *Paramecium*, we shall encounter extremely complicated processes. The corresponding data are given in Table 5 (Appendix) and Figure 26. These experiments were made with Osterhout's medium containing *Bacillus pyocyaneus*, simultaneously with those mentioned above. Therefore, the data on the separate growth of *P. caudatum* and *P. aurelia* given in Appendix Table 3 serve as a control for these experiments.

First of all, the separate growth of *Stylonychia pustulata* is very peculiar: having attained a certain maximum, the density of population decreases and remains stationary at a lower level. Direct observation shows that the bacteria at the close of the twenty-four hour intervals between the changes of medium remain partly unconsumed, and the limiting factor here is apparently an accumulation of waste products and not an insufficiency of food. The fluctuations in the density of population of the separately growing *S. pustulata* are probably connected with some complex processes of the influence of metabolic products on growth. As to the mixed populations, the same regularity with which we had to deal previously repeats itself here: one species finally completely displaces another, and the species displaced is always *Stylonychia pustulata*.

(3) Let us summarize the data of this chapter. We have studied the competition between two species for a source of energy kept continually at a certain level. This process may be divided into two periods. *In the first period the two species compete for the still unutilized resources of energy.* In what proportion this energy will be distributed between the two species is determined by the system of Vito Volterra's differential equations of competition, but the coefficients of the struggle for existence in these equations change in the course of the growth of the population and are therefore more complicated than in the preceding chapter. *In the second period there is but a redistribution of the completely seized energy between the two species,* which is again controlled by the differential equations of competition. Owing to its advantages, mainly a greater value of the coefficient of multiplication, one of the species in a mixed population drives out the other entirely.

Chapter VI

THE DESTRUCTION OF ONE SPECIES BY ANOTHER

I

(1) In the two preceding chapters our attention has been concentrated on the indirect competition, and we have to turn now to an entirely new group of phenomena of the struggle for existence, that of one species being directly devoured by another. The experimental investigation of just this case is particularly interesting in connection with the mathematical theory of the struggle for existence developed on broad lines by Vito Volterra. Mathematical investigations have shown that the process of interaction between the predator and the prey leads to periodic oscillations in numbers of both species, and all this of course ought to be verified under carefully controlled laboratory conditions. At the same time we approach closely in this chapter to the fundamental problems of modern experimental epidemiology, which have been recently discussed from a wide viewpoint by Greenwood in his Herter lectures of 1931. The epidemiologists feel that the spread of microbial infection presents a particular case of the struggle for existence between the bacteria and the organisms they attack, and that the entire problem must pass from the strictly medical to the general biological field.

(2) As the material for investigation we have taken two infusoria of which one, *Didinium nasutum*, devours the other, *Paramecium caudatum* (Fig. 27). Here, therefore, exists the following food chain: bacteria → *Paramecium* → *Didinium*. This case presents a considerable interest from a purely biological viewpoint, and it has more than once been studied in detail (Mast ('09), Reukauf ('30), and others). The amount of food required by *Didinium* is very great and, as Mast has shown, it demands a fresh *Paramecium* every three hours. Observation of the hunting of *Didinium* after the Paramecia has shown that *Didinium* attacks all the objects coming into contact with its seizing organ, and the collision with suitable food is simply due to chance (Calcins '33). Putting it into the words of Jennings ('15) *Didinium* simply "proves all things and holds fast to that which is good."

All the experiments described further on were made with pure lines of *Didinium* ("summer line") and *Paramecium*. In most of the experiments the nutritive medium was the oaten decoction, "with sediment" or "without sediment," described in the preceding chapter. Attempts were also made to cultivate these infusoria on a synthetic medium with an exactly controlled number of bacteria for the Paramecia, but here we encountered great difficulties in connection with differences in the optimal physicochemical conditions for our lines of *Paramecium* and *Didinium*. The introduction of a phosphate buffer and the increase of the alkalinity of the medium above pH =

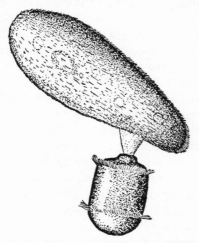

Fig. 27. *Didinium nasutum* devouring *Paramecium caudatum*

6.8–7.0 has invariably favored the growth of *Paramecium*, but hindered that of *Didinium*. Satisfactory results have been obtained on Osterhout's medium, but here also *Didinium* has grown worse than on the oaten medium. Therefore, absolute values of growth under different conditions can not be compared with one another though all the fundamental laws of the struggle for existence remained the same. The experiments were made in a moist thermostat at a temperature of 26°C.

(3) Let us first of all analyze the process of interaction between the predator and the prey from a qualitative point of view. It is well known that under natural conditions periodic oscillations in the numbers of both take place but in connection with the complexity of the

situation it is difficult to draw any reliable conclusions concerning the causes of these oscillations. However, quite recently Lotka (1920) and Volterra (1926) have noted on the basis of a purely mathematical investigation that the properties of a biological system consisting of two species one of which devours the other are such that they lead to periodic oscillations in numbers (see Chapter III). These oscillations should exist when all the external factors are invariable, because they are due to the properties of the biological system itself. The periods of these oscillations are determined by certain initial conditions and coefficients of multiplication of the species. Mathematicians arrived at this conclusion by studying the properties of the differential equation for the predator-prey relations which has already been discussed in detail in Chapter III (equation 21a). Let us now repeat in short this argument in a verbal form. When in a limited microcosm we have a certain number of prey (N_1), and if we introduce predators (N_2),[1] there will begin a decrease in the number of prey and an increase in that of the predators. But as the concentration of the prey diminishes the increase of the predators slows down, and later there even begins a certain dying off of the latter resulting from a lack of food. As a result of this diminution in the number of predators the conditions for the growth of the surviving prey are getting more and more favorable, and their population increases, but then again predators begin to multiply. Such periodic oscillations can continue for a long time. *The analysis of the properties of the corresponding differential equation* shows that one species will *never be capable of completely destroying another:* the diminished prey will not be entirely devoured by the predators, and the starving predators will not die out completely, because when their density is low the prey multiply intensely and in a certain time favorable conditions for hunting them arise. Thus *a population consisting of homogeneous prey and homogeneous predators in a limited microcosm, all the external factors being constant, must according to the predictions of the mathematical theory possess periodic oscillations in the numbers of both species.*[2] These oscillations may be

[1] It is assumed that all individuals of prey and predator are identical in their properties, in other words, we have to do with homogeneous populations.

[2] According to the theory, such oscillations must exist in the case of one component depending on the state of another at the same moment of time, as well as in the case of a certain delay in the responses of one species to the changes of the other.

called "innate periodic oscillations," because they depend on the properties of the predator-prey relations themselves, but besides these under the influence of periodic oscillations of external factors there generally arise "induced periodic oscillations" in numbers depending on these external causes. The classic example of a system which is subject to innate and induced oscillations is presented by the pendulum. Thus the ideal pendulum the equilibrium of which has been disturbed will oscillate owing to the properties of this system during an indefinitely long time, if its motion is not impeded. But in addition to that we may act upon the pendulum by external forces, and thereby cause induced oscillations of the pendulum.

If we are asked what proof there is of the fact tnat the biological system consisting of predator-prey actually possesses "innate" periodic oscillations in numbers of both species, or in other terms that the equation (21a) holds true, we can give but one answer: observations under natural conditions are here of no use, as in the extremely complex natural environment we do not succeed in eliminating "induced" oscillations depending on cyclic changes in climatic factors and on other causes. Investigations under constant and exactly controlled laboratory conditions are here indispensable. Therefore, in experimentation with two species of infusoria one of which devours the other the following question arose at the very beginning: does this system possess "innate" periodic oscillations in numbers, which are to be expected according to the mathematical theory?

(4) The first experiments were set up in small test tubes with 0.5 cm³ of oaten medium (see Chapter V). If we take an oaten medium without sediment, place in it five individuals of *Paramecium caudatum*, and after two days introduce three predators *Didinium nasutum*, we shall have the picture shown in Figure 28. After the predators are put with the Paramecia, the number of the latter begins to decrease, the predators multiply intensely, devouring all the Paramecia, and thereupon perish themselves. This experiment was repeated many times, being sometimes made in a large vessel in which there were many hundreds of thousands of infusoria. The predator was introduced at different moments of the growth of population of the prey, but nevertheless the same result was always produced. Figure 29 gives the curves of the devouring of Paramecia by *Didinium* when the latter are introduced at different moments of the growth of the prey population (in 0.5 cm³ of oaten medium without

sediment). This figure shows the decrease in the number of Paramecia as well as the simultaneous increase in number and in volume
of the population of *Didinium*. (We did not continue these curves
beyond the point where *Didinium* attained its maximal volume.)   It
is evident that the Paramecia are devoured to the very end. As it is
necessary that the nutritive medium should contain a sufficient
quantity of bacteria in order to have an intense multiplication of
Paramecia, we arranged also experiments in the test tubes on a daily
changed Osterhout's medium containing *Bacillus pyocyaneus* (see
Chapter V). In Figure 30 are given the results of such an experiment which has led up, as before, to the complete disappearance of
both *Paramecium* and *Didinium*. Thus we see that in a homogeneous

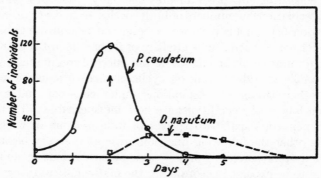

FIG. 28. The elementary interaction between *Didinium nasutum* and *Paramecium caudatum* (oat medium without sediment).   Numbers of individuals
pro 0.5 c.c.   From Gause ('35a).

nutritive medium under constant external conditions the system
Paramecium-Didinium has no innate periodic oscillations in numbers.
In other words, *the food chain:* bacteria → *Paramecium* → *Didinium*
placed in a limited microcosm, with *the concentration of the first link
of the chain kept artificially at a definite level, changes in such a direction that the two latter components disappear entirely and the food resources of the first component of the chain remain without being utilized
by any one.*

We have yet to point out that the study of the properties of the
predator-prey relations must be carried out under conditions favorable for the multiplication of both prey and predator. In our case,
there should be an abundance of bacteria for the multiplication of

Paramecia, and suitable physicochemical conditions for the very sensitive *Didinium*. It is self-evident that if at the very beginning

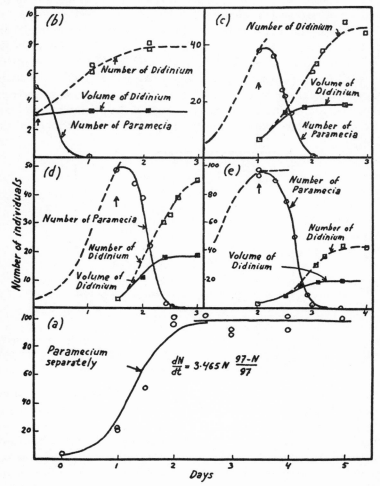

Fig. 29. The destruction of *Paramecium caudatum* by *Didinium nasutum*. (a) Growth of *P. caudatum* alone. (b) *Didinium* is introduced at the very beginning of growth of Paramecia population. (c) *Didinium* is introduced after 24 hours. (d) *Didinium* is introduced after 36 hours. (e) *Didinium* is introduced after 48 hours. Numbers of individuals pro 0.5 c.c.

we set up unfavorable conditions under which *Didinium* begins to degenerate, and as a result is unable to destroy all the prey, or if the diminishing prey should perish not in consequence of their having been devoured by the predators but from other causes, we could not be entitled to draw any conclusions in respect to the properties of the predator-prey relations in the given chain.

(5) We may be told that after we have "snatched" two components out of a complex natural community and placed them under "artificial" conditions, we shall certainly not obtain anything valuable and shall come to absurd conclusions. We will therefore point out beforehand that under such conditions it is nevertheless possible to obtain periodic oscillations in the numbers of the predators and prey, if we but introduce some complications into the arrangement of the

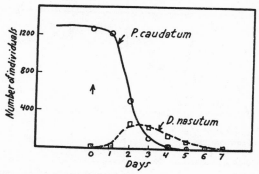

Fig. 30. The elementary interaction between *Didinium nasutum* and *Paramecium caudatum* (medium of Osterhout). The environment is not completely favorable for *Didinium*, and it begins to die out too early. Numbers of individuals pro 5 c.c. From Gause ('35a).

experiments. As yet we have only separated the elementary interaction between two species, and noted some of its fundamental properties.

However, why is the theoretical equation of the mathematicians not realized in our case? The cause of this is apparently that a purely biological property of our predator has not been taken into account in the equation (21a). According to this equation a decrease in the concentration of the prey diminishes the probability of their encounters with the predators, and causes a sharp decrease in the multiplication of the latter, and afterwards this even leads to their partly dying out. However, in the actual case *Didinium* in spite of the insufficiency of food continues to multiply intensely at the expense of a vast decrease in the size of the individual. The following data give an idea of the diminution in size of *Didinium*: three normal individuals of this species placed in a medium free of Paramecia continue to multiply intensely, and in an interval of 24 hours give on an average 7.1 small individuals able to attack the prey. This vast increase of the "seizing surface" represents, metaphorically speaking, those "tentacles by means of which the predators suck out the prey completely." Translating all this into mathematical language, we can say: the function characterizing the consumption of prey by predators $[f_1(N_1, N_2)]$, as well as the natality and the mortality of predators $[F(N_1, N_2)]$,* are apparently more complicated than Lotka and Volterra have assumed in the equation (21a), and as a result the corresponding process of the struggle for existence has no periodic properties. We shall soon return to a further analysis of this problem along mathematical lines.

II

(1) We have but to introduce a slight complication into the conditions of the experiment, and all the characteristic properties of our biological system will be altogether changed. In order to somewhat approach natural conditions we have introduced into the microcosm a "refuge" where Paramecia could cover themselves. For this purpose a dense oaten medium "with sediment" was taken (see Chapter V). Direct observations have shown that while the Paramecia are covered in this sediment they are safe from the attack of predators. It must be noted that the *taxis causing the hiding of Paramecia in this "refuge" manifests itself in a like manner in the presence of the predators as in their absence.*

We must have a clear idea of the rôle which a refuge plays in the struggle for existence of the species under observation, as a lack of clearness can lead further on to serious misunderstandings. If *Didinium* actively pursued a definite *Paramecium* which escaping from it hid in the refuge, the presence of the refuge would be a definite parameter in every elementary case of one species devouring another. In other words, the nature and the distribution of refuges would constitute an integral part of the expressions $f_1(N_1, N_2)$ and $F(N_1, N_2)$ of the corresponding differential equation of the struggle for existence.

---

* See Chapter III, equation (21).

Such a situation has recently been analyzed by Lotka ('32a). We might be told in this case that in experimenting with a homogeneous microcosm *without refuges* we have sharply disturbed the process of elementary interaction of two species. Instead of investigating "in a pure form" the properties of the differential equation of the struggle for existence we obtain a thoroughly unnatural phenomenon, and all the conclusions concerning the absence of innate oscillations in numbers will be entirely unconvincing. But for our case this is not true. We have already mentioned that *Dininium* does not actively hunt for Paramecia but simply seizes everything that comes in its way. In its turn *Paramecium* fights with the predator by throwing out trichocysts and developing an intense rapidity of motion, but *never hiding in this connection in the refuge of our type*. In this manner, we have actually isolated and studied "in a pure form" the elementary phenomenon of interaction between the prey and the predators in a homogeneous microcosm. The refuge in our experiment presents a peculiar "semipermeable membrane," separating off a part of the microcosm into which *Paramecium* can penetrate owing to its taxis, *in general quite independently of any pursuit of the predator*, and which is impenetrable for *Didinium*.

When the microcosm contains a refuge the following picture can be observed (see Fig. 31): if *Paramecium* and *Didinium* are simultaneously introduced into the microcosm, the number of predators increases somewhat and they devour a certain number of Paramecia, but a considerable amount of the prey is in the refuge and the predators cannot attain them. Finally the predators die out entirely owing to the lack of food, and then in the microcosm begins an intense multiplication of the Paramecia (no encystment of *Didinium* has been observed in our experiments). We must make here a technical note: the microcosm under observation ought not to be shaken in any way, as any shock might easily destroy the refuge and cause the Paramecia to fall out. On the whole it may be noted that when there appears a refuge in a microcosm, a certain threshold quantity of the prey cannot be destroyed by the predators. The elementary process of predator-prey interaction goes on to the very end, but the presence of a certain number of undestroyed prey in the refuge creates the possibility of the microcosm becoming later populated by the prey alone.

(2) Having in the experiment with the refuge made the microcosm a heterogeneous one, we have acquired an essential difference of the

corresponding process of the struggle for existence from all the elementary interactions between two species which we have so far examined.   In the case of an elementary interaction between predator and prey in a homogeneous microcosm very similar results were obtained in various analogous experiments (see Table 6, Appendix). In any case the more attention we give to the technique of experimentation, the greater will be this similarity.   In other terms, in a homogeneous microcosm the process of the struggle for existence in every individual test tube was exactly determined by a certain law, and this could be expressed by more or less complex differential equa-

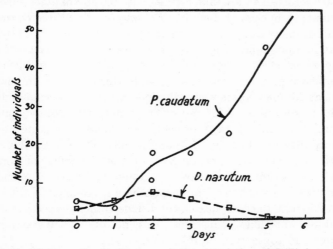

FIG. 31. The growth of mixed population consisting of *Didinium nasutum* and *Paramecium caudatum* (oat medium with sediment).   Numbers of individuals pro 0.5 c.c.

tions.   For every individual microcosm the quantities of the predator and the prey at a certain time *t* could be exactly predicted with a comparatively small probable error.

Such a deterministic process disappears entirely when a refuge is introduced into the microcosm, because the struggle for existence is here affected by a multiplicity of causes.   If we take a group of microcosms with similar initial conditions the following picture is observed after a certain time: (1) in some of the microcosms in spite of the existence of a refuge all the prey are entirely devoured (they might have accidentally left the refuge, hidden inadequately, etc.).

Or else (2) as shown in Figure 31 a certain number of prey might have in the refuge been entirely out of reach of the predators, and the latter will perish finally from lack of food. (3) Lastly, prey may from time to time leave the refuge and be taken by the predators; as a result a mixed population consisting of prey and predators will continue to exist for a certain time. All this depends on the circumstance that in our experiments the absolute numbers of individuals were not large, and the amplitude of fluctuations connected with multiplicity of causes proved to be wider than these numbers.

(3) Let us consider the corresponding data. In one of the experiments 30 microcosms were taken (tubes with 0.5 cm³ of oaten medium with sediment), in each of them five *Paramecium* and three *Didinium* were placed, and two days after the population was counted. It turned out that in four microcosms the predators had entirely destroyed the prey whilst in the other 26 there were predators as well as prey. The number of prey fluctuated from two to thirty-eight. In another experiment 25 microcosms were examined after six days; in eight of them the predators had died out entirely and prey alone remained. Therefore, in the initial stage for every individual microcosm we can only affirm with a probability of $\frac{8}{25}$ that it will develop in the direction indicated in Figure 31. Certain data on the variability of populations in individual microcosms are to be found in Table 7 (Appendix). Further experimental investigations are here necessary. First of all we had to do with too complicated conditions in the microcosms owing to variability of refuges themselves. It is not difficult to standardize this factor and to analyze its rôle more closely.

In concluding let us make the following general remarks. When the microcosm approaches the natural conditions (variable refuges) in its properties, the struggle for existence begins to be controlled by such a multiplicity of causes that we are unable to predict exactly the course of development of each individual microcosm.[3] From the language of rational differential equations we are compelled to pass on to the language of probabilities, and there is no doubt that the corresponding mathematical theory of the struggle for existence may be developed in these terms. The physicists have already had to

---

[3] This means only that the development of each individual microcosm is influenced by a multiplicity of causes, and it would be totally fallacious to conclude that it is not definitely "caused." All our data have of course no relation to the concept of phenomenal indeterminism.

face a similar situation, and it may be of interest to quote their usual remarks on this subject: "Chance does not confine itself here to introducing small, practically vanishing corrections into the course of the phenomenon; it entirely destroys the picture constructed upon the theory and substitutes for it a new one subordinated to laws of its own.   In fact, if at a given moment an extremely small external factor has caused a molecule to deviate very slightly from the way planned for it theoretically, the fate of this molecule will be changed in a most radical manner: our molecule will come on its way across a great number of other molecules which should not encounter it, and at the same time it will elude a series of collisions which should have taken place theoretically.   All these 'occasional' circumstances in their essence are regular and determined, but as they do not enter into our theory they have in respect to it the character of chance" (Chinchin, '29, pp. 164–165).

(4) If we take a microcosm without any refuge wherein an elementary process of interaction between *Paramecium* and *Didinium* is realized, and if we introduce an artificial immigration of both predator and prey at equal intervals of time, there will appear periodic oscillations in the numbers of both species.   Such experiments were made in glass dishes with a flat bottom into which 2 cm³ of nutritive liquid were poured.   The latter consisted of Osterhout's medium with a two-loop concentration of *Bacillus pyocyaneus*, which was changed from time to time.   The observations in every experiment were made on the very same culture, without any interference from without (except immigration) into the composition of its contents.   At the beginning of the experiment and every third day thereafter one *Paramecium* + one *Didinium* were introduced into the microcosm.   The predator was always taken when already considerably diminished in size; if it did not find any prey within the next 12 hours, it usually degenerated and perished.   Figure 32 represents the results of one of the experiments.   Let us note the following peculiarities: (1) At the first immigration into the microcosm containing but few Paramecia the predator did not find any prey and perished.   An intense growth of the prey began.   (2) At the time of the second immigration the concentration of the prey is already rather high, and a growth of the population of the predator begins.   (3) The third immigration took place at the moment of an intense destruction of the prey by the predators, and it did not cause any essential changes.   (4) Towards

the time of the fourth immigration the predator had already devoured all the prey, had become reduced in size and degenerated. The prey introduced into the microcosm originates a new cycle of growth of the prey population. Such periodic changes repeat themselves further on.

Comparing the results of different similar experiments with immigration made in a homogeneous microcosm, we come to the same conclusions as in the preceding paragraph. Within the limits of each cycle when there is a great number of both *Paramecium* and *Didinium* it is possible by means of certain differential equations to predict the course of the process of the struggle for existence for some time to

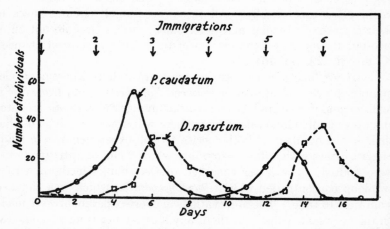

FIG. 32. The interaction between *Didinium nasutum* and *Paramecium caudatum* in a microcosm with immigrations (1 *Didinium* + 1 *Paramecium*). Causes of too low peak of *Didinium* in the first cycle of growth are known. From Gause ('34a).

come. However, at the critical moments, when one cycle of growth succeeds another, the number of individuals being very small, "multiplicity of causes" acquires great significance (compare first and second cycles in Fig. 32). As a result it turns out to be impossible to forecast exactly the development in every individual microcosm and we are again compelled to deal only with the probabilities of change.

(5) Let us briefly sum up the results of the *qualitative analysis* of the process of destruction of one species by another in a case of two infusoria. The data obtained are schematically presented in Figure

33.   In a homogeneous microcosm the process of elementary interaction between the predator and the prey led up to the disappearance
of both the components.   By making the microcosm heterogeneous

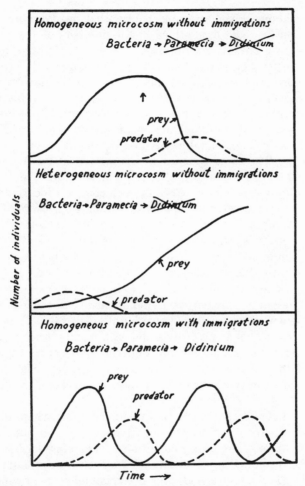

Fig. 33. A schematic representation of the results of a qualitative analysis
of the predator-prey relations in the case of two Infusoria.

(refuge) and thus approaching the natural conditions we began to
deal with a "probability" of change in various directions.   The predator sometimes dies out and only prey populate the microcosm.   By

introducing immigration into a homogeneous microcosm we obtain periodic oscillations in the numbers of both species.

III

(1) The above given example shows that in *Paramecium* and *Didinium* the periodic oscillations in the numbers of the predators and of the prey are not a property of the predator-prey interaction itself, as the mathematicians suspected, but apparently occur as a result of constant interferences from without in the development of this interaction.   There is evidence for believing that this is characteristic for more than our special case.   Jensen ('33) in his monograph on periodic fluctuations in size of various stocks of fish concluded that he has not found any periodicity identic to those treated by Volterra.   The same conclusion was arrived at by S. Severtzov ('33) dealing with vertebrates.   There are also plenty of entomological observations showing the possibility of a complete local extermination of hosts by parasites.   We may according to Cockerell ('34) recall some observations on Coccidae (scale insects) made in New Mexico.   Certain species occur on the mesquite and other shrubs which exist in great abundance over many thousands of square miles of country.   Yet the coccids are only found in isolated patches here and there.   They are destroyed by their natural enemies, but the young larvae can be blown by the wind or carried on the feet of birds, and so start new colonies which flourish until discovered by predators and parasites.   This game of hide-and-seek results in frequent local exterminations, but the species are sufficiently widespread to survive in parts of their range, and so continue indefinitely.   Such local exterminations in grayfish have been recently observed by Duffield ('33).

(2) Experimental epidemiology is the one domain where the problems of a direct struggle for existence have already been submitted to an exact analysis in laboratory conditions.   Therefore let us consider in brief the results there obtained.   If we took a microcosm of any size populated by homogeneous organisms, not allowing any immigration or emigration, and if we caused it to be fatally infected, we should obtain a complete dying out of the organisms (if among them there were no immune ones, and if they were unable to acquire any immunity).   In other terms, we should have before us the well-known elementary interaction of two species.   However, the process of dying

out does not usually go on to the end owing to the heterogeneity of population and presence of immune individuals, which are in a certain degree equivalent to the individuals protected in the refuge: they are not carried away by the process of destruction which goes on to the end among the non-immune ones. The nature of this "refuge" is very complicated, and it is interesting to quote here the following words of Topley ('26, pp. 531–532): "Most of us who have been concerned at all with the problem of immunity have been accustomed to take the individual as our unit. When we take as our unit not the individual but the herd, entirely new factors are introduced. Herd-resistance must be studied as a problem *sui generis*. One factor peculiar to the development of communal as opposed to individual immunity may be referred to here. A herd may clearly increase its average resistance by a process of simple selection, by the elimination through death of its more susceptible members. . . . That some process of active immunization will be associated with the occurrence of non-fatal infection may safely be assumed, though its degree and importance may be very difficult to assess, so that we must allow for the possibility that the spread of infection which is killing some of our hosts is immunizing others.

"A very imperfect analogy may help to depict the position. Suppose we take a number of stakes of different thickness, plant them in the ground and expose them to bombardment with stones of varying size from catapults of varying strengths. After a certain time we shall find that a number of the stakes have been broken. This will not have happened to many of the thicker stakes, but other survivors will consist of thinner stakes, around which ineffective missiles have formed a protective armour. Survivors of the latter class are in a precarious state; subsequent bombardment may displace the protective heap, and perhaps add its impetus to that of the new missile. Survivors of the former class may eventually be destroyed by a missile of sufficient momentum."

(3) The experiments of epidemiologists dealing with the influence of immigration on the course of an epidemic among mice in a limited microcosm are also very interesting. One can distinctly see here that a continuous interference from without acting upon a definite population causes periodic oscillations in the epidemic which disappear immediately as the interference ceases. Let us quote Topley again: "When susceptible mice gain access to the cage at a steady

rate the deaths are not uniformly distributed in time, nor do they occur in a purely random fashion. They are grouped in a series of waves, each wave showing minor fluctuation. The equilibrium between parasite and host seems to be a shifting one. As the result of some series of changes, the parasite appears to obtain a temporary mastery, so that a considerable proportion of the mice at risk fall victims to a fatal infection. This is followed by a phase in which there is a decreased tendency for the occurrence of fatal infection, and the death-rate falls. As fresh susceptibles accumulate this succession of events is repeated, and the deaths increase to a fresh maximum, only to fall again when this maximum is passed." But if only "no such immigration occur the epidemic gradually dies down, leaving a varying number of survivors."

We can conclude that *the process of elementary interaction between the homogeneous hosts and the homogeneous bacterial population possesses no "classical" periodic variations.* Without wishing to adopt at once the preconceived opinion that such a phenomenon is generally impossible, we ought in any case require a clear demonstration of its possibility. This demonstration will be really given below.

IV

(1) Turning back again from empirical observations to the general principles let us note that there can exist two different types of innate periodic oscillations in the systems, as Hill ('33) has noted recently in connection with physiological problems. One of them which was assumed by Lotka-Volterra and which we have searched above must be called a "classical" fluctuation and it is entirely analogous to well-known oscillations in physics arising as the consequence of the reaction with one another of properties analogous to inertia and elasticity. A changing system tends, on one hand, to maintain its state of motion because it possesses mass, whilst on the other the force of elasticity increases according to the removal from equilibrium and ultimately reverses the motion or change. In the classical theory of biological population the predator tends to multiply indefinitely, but by a removal in this way from an equilibrium with the prey the change in the predator population becomes reversed, later again replaced by an increase, and so on (equation 21a).

There is, however, another type of oscillation with which physiologists are concerned and to which apparently belong the spread of epidemics and fluctuations in our protozoan population. A certain potential or a certain state is here built up by a continuous process and the conditions become less and less stable until a state is reached at which a discharge (or epidemics) must take place. It is evident that the interaction between the two components instead of periodicity leads here to an interruption of contact (depending from specific biological conditions in the case of epidemics and from a disappearance of predators and prey in our Protozoa), and then ceases until the next critical threshold. Such oscillations with an interruption of contact bear in physics the name of "relaxation oscillations."

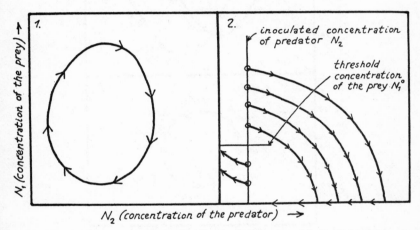

Fig. 34. Diagrams illustrating two types of innate periodic fluctuations in numbers of animals. 1. "Classical" fluctuation of Lotka-Volterra. 2. "Relaxation fluctuations."

It is easy to visualize the difference between these two types of oscillations employing the illuminating graphs so often used by Lotka. On the coördinate paper we usually plot time on the abscissae and densities of predators (N_2) and prey (N_1) on the ordinates. But if we abstract from time and plot N_2 on the abscissae and N_1 on the ordinates we obtain a clear idea of the nature of interspecific interaction. As Figure 34 shows in the case of classical oscillation we must have a closed curve.[4]

[4] The transformation of the usual time-curves into such graphs is illustrated by a numerical example reproduced in Fig. 35. The upper part of it presents a

Let us now turn our attention to the graph for the relaxation inter-action. Suppose we introduce a definite amount of the predator, N_2, at different densities of the prey (N_1). Then, before the critical threshold of the latter is reached ($N_1°$, Fig. 34), an epidemic of *Didinium* cannot start and the curves return on the ordinate. After the

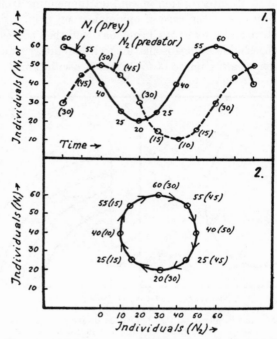

FIG. 35. Diagram illustrating the transformation of the usual time-curves (1) into relative graphs of interaction (2) for the "classical" Lotka-Volterra fluctuation in numbers.

critical threshold is reached there appears a relaxation which leads to the destruction of the prey—the curves cross the abscissa.

(2) This little amount of theory enables us to formulate our problem thus: How do the biological adaptations consisting of a very active consumption of *Paramecium* by *Didinium* disturb the condi-

theoretical case of the classical Volterra's oscillation in the usual form. If we note the values of N_1 and N_2 at different moments of time, and then plot N_1 against the corresponding N_2, we shall obtain the closed curve reproduced below.

tions of the classical equation (21a) and transform it into that of an elementary relaxation? For all the technical details the reader is referred to the original paper (Gause and Witt, '35), and we will discuss here only its essential ideas.

In a first approximation to the actual state of affairs we can write an elementary equation of relaxation. It can be admitted [on the basis of the observations on *Paramecium* (N_1) and *Didinium* (N_2)] that if N_2 is large the mortality of the predators is negligible when $N_1 > 0$. In addition, the increase of predators only slightly depends on N_1 (with an insufficiency of prey the predators continue to multiply at the expense of a decrease in size of the individuals; in this connection the consumption of prey but slightly depends on N_1).

Introducing these assumptions into the equation (21) we write $-\dfrac{dN_2}{dt} = 0$ where $N_1 \neq 0$ and $-\dfrac{dN_2}{dt} = d_2N_2$ where $N_1 = 0$.* To reduce the dependence upon N_1 of the members characterizing the interaction of species, we substitute $\sqrt{N_1}$ to N_1.† Then

$$\left.\begin{aligned}
\frac{dN_1}{dt} &= b_1N_1 - k_1N_2\sqrt{N_1} \\
\frac{dN_2}{dt} &= b_2N_2\sqrt{N_1} \qquad\qquad (N_1 \neq 0) \\
&= -d_2N_2 \qquad\qquad\quad (N_1 = 0)
\end{aligned}\right\} \cdots (21b)$$

Figure 36 shows that the solution of the equation 21b (the integral curves on the graph N_1, N_2) actually coincides with biological observations on *Paramecium* and *Didinium*. It is therefore safe to assume that the general equation of the destruction of one species by another (21) takes in our special case the form (21b) instead of the classical expression of Lotka-Volterra (21a).

(3) The equation of relaxation (21 b) represents but a first approximation to the actual state of things, and is true only if N_1 or N_2 are large. Looking at the trend of the experimental curves on the surface N_1, N_2 with small densities (Fig. 37) we notice that they pass from the right to the left and cross the ordinate (Fig. 37, a). This means

* This condition is already sufficient for an exclusion of the "classical" periodic fluctuations.

† Special experiments show that this substitution is satisfactory (Gause and Witt, '35).

that "an epidemic" of predators cannot break out if the concentration of the prey has not attained the threshold value a_h. Below it predators disappear[5] and leave a pure population of prey, but above it we find usual relaxations.

Taking into account all these features we can write for *Paramecium* and *Didinium* a complicated equation of relaxation representing an adequate expression of what actually exists. We admit that the mortality of predators appears not only with $N_1 = 0$, but that a slight mortality generally exists increasing with a diminution of the concen-

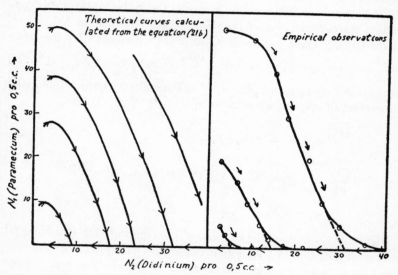

FIG. 36. The solution of the equation 21b (to the left) and empirical observations on *Paramecium* and *Didinium* (to the right). No "residual growth" of the population of predators (in the absence of the prey) is taken into account in the theoretical equation. From Gause and Witt, '35.

tration of N_2, and that the intensity of hunting also increases with an insufficiency of prey.

[5] In these experiments were used predators possessing no 'residual growth,' e.g. already diminished in size. See experiments with immigrations where predators usually did not find any prey in the microcosms containing very few Paramecia, and consequently perished.

The solution of this complicated equation[6] is represented on Figure 38. It is a further concretization for *Paramecium* and *Didinium* of the principle of relaxation represented on Figure 34. An epidemic

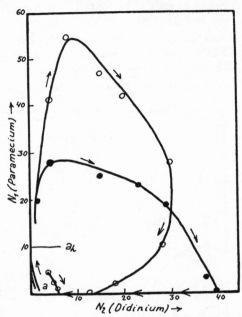

FIG. 37. The interaction between *Paramecium* and *Didinium* at different densities of population. The absence of the "residual growth" (comp. Fig. 36, right) as well as the differences between both curves are connected with slightly unfavorable conditions of the medium.

of predators cannot start below the threshold in the concentration of the prey, but above it we find usual relaxations. A characteristic feature of our food-chain is an extraordinarily low value of the threshold.

[6] The equation given by Gause and Witt ('35) is:

$$
\left.
\begin{aligned}
\frac{dN_1}{dt} &= b_1 N_1 - f_1(N_1)\, N_1 N_2 \\[2mm]
\frac{dN_2}{dt} &= b_2 N_2 \sqrt{N_1} - f(N_2) \qquad N_1 \neq 0 \\[2mm]
&= -\, d_2 N_2 \qquad\qquad\qquad N_1 = 0
\end{aligned}
\right\} \quad\dots\dots\dots\dots (21c)
$$

A mathematical analysis of the properties of the complicated equation of relaxation given in Fig. 38 shows that there is a point on the map of the curves of interaction which is usually called "singular point." The powers of mortality, natality and interaction of prey and predators are so balanced that the "classical" oscillations in numbers are theoretically possible around it. But in the case of *Paramecium* and *Didinium* the coördinates of this singular point are exceedingly small. In other terms the zone of possible classical oscillations is displaced here to such small densities that these oscillations

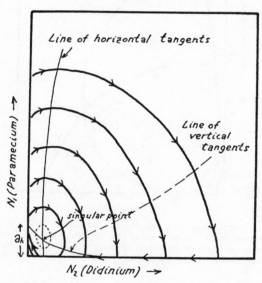

FIG. 38. The solution of the complicated equation of relaxation (21c). a_h represents the threshold concentration of the prey.

are completely annihilated by the statistical factors which are much more powerful in this zone.

(4) In conclusion let us consider the appearance of periodic variations in numbers under the influence of immigrations (a slight and synchronous inflow of N_1 and N_2 after intervals of time t). In other terms we have to deal here with the problem of the influence of small impulses. At the origin (a, Fig. 37) they lead to a return of the curve to the ordinate. Relaxations arise when the concentration of N_1 rises above the threshold. From Figure 38 it is easy to calculate how a

delay of the inflow after the threshold has been reached increases the
dimensions of the relaxations (the importance of this problem for
epidemiology has been pointed out by Kermack and McKendrick
('27)). When relaxation is going on, slight impulses do not disturb it
seriously until crossing of the abscissa by the integral curves, and later
on up to their intersection with the line of horizontal tangents (Fig.
38). After this the impulses lead to a return on the ordinate and the
process begins again.

V

(1) The theory of the preceding section shows that the consumption of one species by another in the population studied is so active
that the classical oscillations in numbers are transformed into an
elementary relaxation and the coördinates of the singular point
around which such oscillations could be theoretically expected are
exceedingly small. This fact except its independent interest enables
us to predict that were we in a position to reduce the intensity of
consumption we could increase the coördinates of the singular point,
and in this way observe the classical oscillations of Lotka-Volterra.
The situation is entirely analogous to that of classical physiology.
The rate of propagation of nervous impulses is under usual conditions
too high and it is sometimes desirable to decrease it, to cool the nerve,
in order to be able to observe certain phenomena. How can one decrease the intensity of consumption of one species by another?

The simplest way is to investigate a system where this intensity is
naturally low. This has been recently made by Gause ('35b) who
analyzed the properties of the food chain consisting of *Paramecium
bursaria* and *Paramecium aurelia* devouring small yeast cells, *Schizosaccharomyces pombe* and *Saccharomyces exiguus*. Special arrangements allowed of controlling artificially the mortality of predators by
rarefying them, and of avoiding the settling of yeast cells on the
bottom by a slow mixing of the medium. Figure 39 shows that under
such specialized conditions fluctuations of the Lotka-Volterra type
actually take place, and in this manner the conditions of the equation
(21a) are realized in general features.

It must be remarked that the equation (21a) does not hold absolutely true because the oscillations do not apparently belong to the
"conservative" type. In other terms they do not keep the magnitude
initially given them but tend to an inherent magnitude of their own

(compare the first and second cycles on Figure 39, ₁ and ₂). This problem, however, requires further investigations.

(2) It is interesting to compare our data with some observations

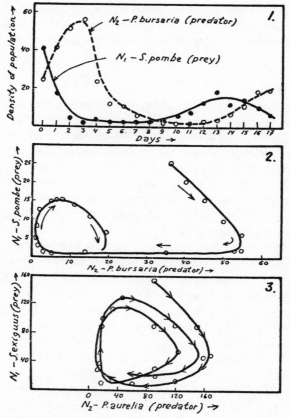

Fig. 39. "Classical" periodic fluctuations in a mixed population of *Paramecium bursaria* and *Schizosaccharomyces pombe* in the usual (1) and in the relative form (2). (3) Periodic fluctuations in a population of *Paramecium aurelia* and *Saccharomyces exiguus* in the relative form. According to Gause, '35b.

made at the Rothamsted Station (Cutler, '23, Russell, '27) which also attracted the attention of Nikolson ('33). There is reason to suspect that the fluctuations in the numbers of soil organisms observed by Cutler and referred to by Russell are interspecific oscillations. "Bacteria do not fluctuate in numbers when grown by themselves in sterilized soil; they rise to high numbers and remain at approximately a constant level. Their numbers fall, however, as soon as the soil amoebae are introduced, but no constant level is reached; instead there are continuous fluctuations as in normal soils. There is a sharp inverse relationship between the numbers of bacteria and those of active amoebae; when the numbers of amoebae rise, those of bacteria fall." It is hardly possible therefore to avoid the conclusion that Cutler had to deal here (in a complicated form) with classical periodic variations of the Volterra type.

In conclusion let us note that this final demonstration of the possibility of "classical" oscillations showed that very specialized conditions are required for their realization, and it is therefore easy to understand why in real biological systems with their typical adaptations leading to very intensive attacks of one species on another the so much discussed "relaxation interaction" between the species apparently predominates.

VI

(1) In Chapter III we have pointed out that the connection between the relative increase of the predator and the number of prey is not a linear one, and that this is of significance for the processes of one species devouring another. We can now be convinced that this connection is actually non-linear.

Recently Smirnov and Wladimirow ('34) have investigated under laboratory conditions the connection between the density of the hosts N_1 (pupae of the fly *Phormia groenlandica*) and the relative increase of the parasite $\frac{1}{N_2} \frac{dN_2}{dt}$ [the progeny of one pair ($\male + \female$) of a parasitic wasp, *Mormoniella vitripennis*, per generation]. The experimental data they obtained are represented in Figure 40. As the density of the hosts increases, the relative increase of the parasite increases also until it reaches the maximal possible or "potential increase" (b_2) from one pair under given conditions. That the curve showing the connection between $\frac{1}{N_2} \frac{dN_2}{dt}$ and N_1 can actually be expressed by the equation [Chapter III, (23)]:

$$\frac{1}{N_2} \frac{dN_2}{dt} = b_2 \left(1 - e^{-\lambda N_1}\right)$$

can be seen in the following manner: by plotting as ordinates the values $\log\left(b_2 - \dfrac{1}{N_2}\dfrac{dN_2}{dt}\right)$ corresponding to different values of abscissae N_1, we should obtain a straight line which, as Figure 40 shows, is actually observed. The slope of this straight line is characterized by the coefficient λ, which thus expresses the rate at which the relative increase of the parasites approaches its maximal value with the increase of the density of hosts.

(2) To summarize: We expected at the beginning of this chapter to find "classical" oscillations in numbers arising in consequence of the continuous interaction between predators and prey as was assumed

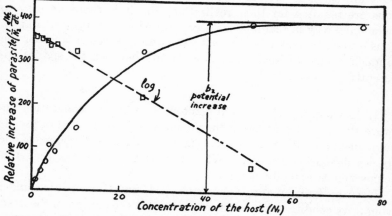

Fig. 40. Connection between the progeny of one pair of the parasite, *Mormoniella vitripennis*, and density of the host, *Phormia groenlandica*, according to Smirnov and Wladimirow. From Gause, '34c.

by Lotka and by Volterra. But it immediately became apparent that such fluctuations are impossible in the population studied, and that this holds true for more than our special case. The corresponding analysis showed definitely to what biological adaptations this impossibility is due. This has enabled us to find a particular system possessing no such adaptations and in this way to observe "classical" fluctuations under very specialized conditions.

It is to be hoped that further experimental researches will enable us to penetrate deeper into the nature of the processes of the struggle for existence. But in this direction many and varied difficulties will undoubtedly be encountered.

APPENDICES

TABLE 1

The growth of the yeast volume and the number of cells in pure cultures of Saccharomyces cerevisiae, Schizosaccharomyces kephir and in the mixed population of these species

This table is taken from Gause (32b)

AGE IN HOURS	SACCHAROMYCES			MIXED POPULATION				SCHIZOSACCHAROMYCES			NUMBER OF CELLS PER UNIT OF YEAST VOLUME		VOLUMES OF THE SPECIES IN THE MIXED POPULATION ESTIMATED FROM THE NUMBER OF CELLS	
	Volume of yeast	Number of squares counted	Average number of cells per square	Volume of yeast	Number of squares counted	Average number of cells per square Sa.	Schi.	Volume of yeast	Number of squares counted	Average number of cells per square	Sa.	Schi.	Sa.	Schi.
Experiment 1														
6	0.37	50	6.24	0.50	50	6.22	16.82	—	—	—	16.86	—	0.375	0.291
16	8.87	60	174.2	6.83	60	66.2	56.4	1.00	50	61.35	19.64	61.35	3.99	0.98
24	10.66	50	166.7	8.66	50	77.8	84.8	1.70	50	88.70	15.63	52.18	4.69	1.47
29	12.50	50	219.2	9.07	50	102.4	84.2	2.73	50	162.60	17.53	59.58	6.15	1.46
40	13.27	50	202.0	—	—	—	—	4.87	—	—	15.22	—	—	—
48	12.87	50	188.6	10.23	50	120.7	98.9	5.67	—	—	14.61	—	7.27	1.71
53	12.70	—	—	10.50	50	137.6	106.2	5.80	—	—	—	—	8.30	1.84
72	—	—	—	—	—	—	—	5.83	—	—	—	—	—	—
93	—	—	—	—	—	—	—	—	—	—	—	—	—	—
117	—	—	—	—	—	—	—	—	—	—	—	—	—	—
141	—	—	—	—	—	—	—	—	—	—	—	—	—	—
Mean											= 16.59	= 57.70		
Experiment 2														
7.5	1.63	60	25.72	1.33	60	16.54	20.06	—	—	—	15.78	—	0.923	0.371
15.0	6.20	60	99.75	4.80	60	55.20	34.08	1.27	60	61.41	16.09	48.38	3.082	0.630
24.0	10.97	60	203.50	8.57	60	103.5	65.9	—	—	—	18.55	—	5.780	1.220
31.5	12.60	60	243.60	11.17	60	177.6	60.1	2.33	60	126.3	19.33	54.20	9.910	1.112
33.0	12.90	60	244.70	10.80	60	169.6	66.2	—	—	—	18.96	—	9.470	1.225
44.0	12.77	60	248.60	11.47	60	189.3	59.5	—	—	—	19.46	—	10.570	1.102
51.5	12.90	60	223.00	10.80	60	177.1	51.8	4.56	60	272.0	17.28	59.65	9.883	0.961
Mean											= 17.92	= 54.08		

TABLE 2

Alcohol production in Saccharomyces cerevisiae and Schizosaccharomyces kephir under aerobic and anaerobic conditions (1932)

AGE IN HOURS	SACCHAROMYCES ANAEROBIC		SACCHAROMYCES AEROBIC		AGE IN HOURS	SCHIZOSACCHARO-MYCES ANAEROBIC		SCHIZOSACCHARO-MYCES AEROBIC	
	Alcohol, per cent	Alcohol per unit of yeast volume	Alcohol, per cent	Alcohol per unit of yeast volume		Alcohol, per cent	Alcohol per unit of yeast volume	Alcohol, per cent	Alcohol per unit of yeast volume
19	1.04	0.201	1.97	0.225	19	—	—	1.28	0.267
15.5	0.45	0.269	1.23	0.167	32	0.51	0.490	1.89	0.278
24.5	1.37	0.264	1.93	0.225	53	0.72	0.487	—	—
19	—	—	2.05	0.211	69	0.82	0.482	—	—
					15.5	—	—	0.81	0.176
					34	0.58	0.523	1.78	0.291
					34	0.51	0.510	1.86	0.305
					53.5	0.93	0.547	1.56	0.231
					53.5	0.78	0.497	—	—
					70.5	1.03	0.545	—	—
		Mean = 0.245		Mean = 0.207			Mean = 0.510		Mean = 0.258

$$\alpha_1 = \frac{0.510}{0.245} = 2.08 \qquad \alpha_2 = \frac{0.258}{0.207} = 1.25$$

TABLE 3

The growth of the number of individuals of Paramecium caudatum and Paramecium aurelia in experiments with the medium of Osterhout

AGE IN DAYS	P. AURELIA				P. CAUDATUM					MIXED POPULATION						Mean	
	Number of individuals per 0.5 c.c. in the culture No.:			Mean	Number of individuals per 0.5 c.c. in the culture No.:				Mean	No. 1		No. 2		No. 3			
	1	2	3		1	2	3	4		P.a.	P.c.	P.a.	P.c.	P.a.	P.c.	P.a.	P.c.
0	2	2	2	2	2	2	2	2	2	2	2	2	2	2	2	2	2
1	—	—	—	—	—	—	—	—	—	—	—	—	—	—	—	—	—
2	17	15	11	14	13	8	13	6	10	7	7	6	9	16	13	10	10
3	29	36	37	34	19	9	6	7	10	25	14	13	5	25	15	21	11
4	39	62	67	56	16	14	7	6	11	43	26	41	41	89	19	58	29
5	63	84	134	94	31	16	16	22	21	81	65	74	60	120	26	92	50
6	185	156	226	189	97	57	40	32	56	140	72	195	135	270	57	202	88
7	258	234	306	266	140	94	98	84	104	180	126	164	92	144	88	163	102
8	267	348	376	330	180	142	124	100	137	224	136	160	148	280	88	221	124
9	392	370	485	416	204	175	145	138	165	240	120	240	90	400	70	293	93
10	510	480	530	507	264	150	175	189	194	204	105	230	90	275	45	236	80
11	570	520	650	580	220	200	280	170	217	375	65	215	65	320	67	303	66
12	650	575	605	610	180	172	240	204	199	370	115	230	50	305	85	302	83
13	560	400	580	513	175	189	230	210	201	285	60	285	60	450	45	340	55

TABLE 3—*Concluded*

AGE IN DAYS	P. AURELIA Number of individuals per 0.5 c.c. in the culture No.: 1	2	3	Mean	P. CAUDATUM Number of individuals per 0.5 c.c. in the culture No.: 1	2	3	4	Mean	MIXED POPULATION Number of individuals per 0.5 c.c. in the culture No.: 1 P.a.	1 P.c.	2 P.a.	2 P.c.	3 P.a.	3 P.c.	Mean P.a.	Mean P.c.
14	575	545	660	593		234	171	140	182	400	65	360	65	402	72	387	67
15	650	560	460	557		192	219	165	192	325	50	275	40	405	65	335	52
16	550	480	650	560		168	216	152	179	370	60	350	40	370	65	363	55
17	480	510	575	522		240	195	135	190	320	40	350	60	300	20	323	40
18	520	650	525	565		183	216	219	206	300	45	405	65	370	35	358	48
19	500	500	550	517		219	189	219	209	315	60	330	40	280	40	308	47
20				500					196							350	50
21				585					195							330	40
22				500					234							350	20
23				495					210							350	20
24				525					210							330	35
25				510					180							350	20

TABLE 4

The growth of the number of individuals of Paramecium caudatum and Paramecium aurelia in experiments with the buffered medium

"One loop" and "half loop" concentrations of bacteria

AGE IN DAYS	NUMBER OF P. AURELIA PER 0.5 C.C. One loop	Half loop	NUMBER OF P. CAUDATUM PER 0.5 C.C. One loop	Half loop	NUMBER OF INDIVIDUALS PER 0.5 C.C. IN THE MIXED POPULATION One loop P.a.	One loop P.c.	Half loop P.a.	Half loop P.c.
0	2	2	2	2	2	2	2	2
1	6	3	6	5	10	5	4	8
2	24	29	31	22	29	15	29	20
3	75	92	46	16	68	32	66	25
4	182	173	76	39	144	52	141	24
5	264	210	115	52	164	40	162	—
6	318	210	118	54	168	32	219	—
7	373	240	140	47	248	36	153	—
8	396	—	125	50	240	40	162	21
9	443	—	137	76	—	32	150	15
10	454	240	162	69	281	20	175	12
11	420	219	124	51	—	30	260	9
12	438	255	135	57	300	12	276	12
13	492	252	133	70	—	16	285	6
14	468	270	110	53	—	20	225	9
15	400	240	113	59	360	12	222	3
16	472	249	127	57	294	9	220	0

TABLE 5

The growth of the number of individuals of Stylonychia pustulata, Paramecium caudatum and Paramecium aurelia in experiments with the medium of Osterhout

S. PUSTULATA — Number of individuals per 0.5 c.c. in the culture No.:

Age in days	1	2	3	4	5	Mean
0	2	2	2	2	2	2
1	7	2	13	26	11	12
2	18	14	38	59	57	37
3	39	59	64	93	73	66
4	47	92	74	90	68	74
5	66	97	70	70	77	76
6	72	78	64	55	40	62
7	51	64	20	34	28	39
8	51	62	41	39	37	45
9	46	64	34	35	45	50
10	74	48	21	37	34	41
11	63	33	34	43	38	39
12	49	30	24	41	42	38
13	54	43	35	47	44	39
14	27	45	22	34	38	38
15	51	25	30	40	36	33
16	36	55	30	31	47	41
17		57	28	31	55	43
18		52	44	62	66	56
19						54
20						34
21						32
22						36
23						40
24						41

MIXED POPULATION OF S. PUSTULATA AND P. AURELIA — Number of individuals per 0.5 c.c. in the culture No.:

Age in days	1 P.a.	1 S.p.	2 P.a.	2 S.p.	3 P.a.	3 S.p.	Mean P.a.	Mean S.p.
0	2	2	2	2	2	2	2	2
1	9	1	15	16	19	14	14	10
2	12	10	41	32	39	46	31	29
3	76	23	50	40	52	50	59	38
4	72	50	120	64	150	56	114	57
5	180	45	270	75	264	57	238	59
6	249	51	275	40	260	28	261	40
7	400	48	316	36	332	42	349	42
8	370	40	440	30	450	35	420	35
9	445	40	410	20	450	25	435	28
10	525	20	455	20	490	5	490	15
11	505	15	460	20	470	20	478	17
12	550	25	475	5	375	20	467	17
13	390	10	430	5	405	15	408	10
14	535	25	410	10	350	0	432	12
15	505	10	385	5	415	15	435	10
16	600	25	420	5	400	5	475	12
17	400	20	440	0	440	5	427	8
18	550	5	395	0	395	5	447	2
19							360	0
20							475	0
21							500	5
22							450	0
23							470	0
24							470	0

MIXED POPULATION OF S. PUSTULATA AND P. CAUDATUM — Number of individuals per 0.5 c.c. in the culture No.:

Age in days	1 P.c.	1 S.p.	2 P.c.	2 S.p.	3 P.c.	3 S.p.	Mean P.c.	Mean S.p.
0	2	2	2	2	2	2	2	2
1	7	12	7	11	4	14	6	12
2	13	56	17	49	8	42	13	49
3	30	76	22	60	28	34	27	57
4	44	54	40	66	46	94	43	71
5	50	48	80	32	60	70	63	50
6	102	30	105	38	102	60	103	43
7	120	40	108	36	128	40	119	39
8	118	24	90	36	108	52	105	37
9	171	30	118	63	135	30	141	41
10	120	30	130	33	160	24	137	29
11	131	51	117	42	150	30	133	41
12	115	33	102	33	108	36	108	34
13	165	36	110	39	100	18	125	31
14	174	24	132	21	147	18	151	21
15	200	24	111	15	153	15	155	18
16	150	6	110	30	120	15	127	17
17	174	9	141	6	126	9	147	8
18	180	18	150	30	108	0	146	16
19							150	6
20							180	12
21							200	15
22							225	9
23							198	12
24							170	3

TABLE 6

The growth of a mixed population consisting of Didinium nasutum and Paramecium caudatum in the medium without sediment

Variations of individual cultures

Number of individuals of $\dfrac{P.c.}{D.n.}$ in the tube with 0.5 c.c. of the medium.   Each group consists of tubes of the same age

$\frac{18}{3}$	$\frac{13}{6}$	$\frac{10}{7}$	$\frac{11}{8}$	$\frac{8}{8}$	$\frac{44}{13}$	$\frac{47}{10}$	$\frac{23}{10}$	$\frac{25}{9}$	$\frac{56}{9}$
$\frac{0}{19}$	$\frac{0}{11}$	$\frac{0}{19}$	$\frac{0}{15}$	$\frac{0}{18}$	$\frac{1}{31}$	$\frac{0}{31}$	$\frac{0}{38}$	$\frac{0}{39}$	$\frac{0}{24}$
$\frac{0}{32}$	$\frac{0}{27}$	$\frac{0}{23}$	$\frac{0}{19}$	$\frac{0}{22}$	$\frac{0}{41}$	$\frac{0}{47}$	$\frac{0}{53}$	$\frac{0}{45}$	$\frac{0}{40}$

TABLE 7

The growth of a mixed population consisting of Didinium nasutum and Paramecium caudatum in the medium with sediment

Variations of individual cultures

Number of individuals of $\dfrac{P.c.}{D.n.}$ in the tube with 0.5 c.c of the medium.   Each group consists of tubes of the same age.

$\frac{48}{2}$	$\frac{20}{7}$	$\frac{31}{6}$	$\frac{55}{0}$	$\frac{72}{0}$	$\frac{3}{0}$	$\frac{24}{3}$	$\frac{20}{4}$	$\frac{7}{12}$	$\frac{46}{7}$

II. APPENDIX TO CHAPTER IV

CALCULATION OF THE COEFFICIENTS OF MULTIPLICATION AND OF THE EQUATIONS OF THE STRUGGLE FOR EXISTENCE

We have had to deal frequently with the coefficients of geometric increase or coefficients of multiplication as well as with the equations of competition. For calculating the latter a very simple approximate integration by the method of Runge-Kutta is necessary, with which, however, most biologists are not familiar. Therefore we will briefly give here the technical details of the calculations and examine a concrete example.

1. We possess experimental data on the separate growth of the first and second species. We wish to calculate the maximal masses, K_1 and K_2, and the coefficients of multiplication, b_1 and b_2.

Example: The growth of *Schizosaccharomyces kephir* under anaerobic conditions (1931). Placing the experimental data on the graph we determine approximately the maximal mass K (Fig. 41). For calculation of the parameter b let us apply the method elaborated by Pearl and Reed. For every individual observation (N_1, N_2, etc.) we take $K - N$, and then $\dfrac{K - N}{N}$ and log $\dfrac{K - N}{N}$. Placing the calculated log $\dfrac{K - N}{N}$ against the corresponding values of time (t) we should obtain a straight line in the case of a symmetrical S-shaped curve (Fig. 41, bottom). We draw this straight line approximately, approaching the experimental observations.

Some brief theoretical explanations are needed here. The differential equation of Pearl's logistic curve: $\dfrac{dN}{dt} = bN \dfrac{K - N}{K}$ after integration takes this form: $N = \dfrac{K}{1 + e^{a-bt}}$. Hence it follows that $e^{a-bt} = \dfrac{K - N}{N}$, and therefore $\log \dfrac{K - N}{N} = (a - bt) \log e$. Taking natural logarithms we obtain: $ln \dfrac{K - N}{N} = a - bt$. Now it is clear that if we place the Napierian logarithms of $\dfrac{K - N}{N}$ against the absolute values of time (t) we must obtain a straight line whose tangent will be the required coefficient of multiplication b, whilst a characterizes the position of the zero ordinate. Their calculation is reduced to determining the equation of the straight line we have drawn approximately. For this purpose we measure the ordinates of two points of the straight line, for instance $t = 0$, $\log_{10} y = m$, and $t = 80$, $\log_{10} y = n$. Passing on to the natural logarithms, i.e. multiplying m by 2.3026 we obtain directly the coefficient a ($a - 0b = m \times 2.3026$). We can then easily calculate b in this way: $a - 80b =$

$n \times 2.3026$.   In order to verify the calculations it is necessary to insert all the parameters into the logistic equation: $N = \dfrac{K}{1 + e^{a-bt}}$ and by substituting different values in place of t to find the "calculated curve," which must approach closely the observations.   A more detailed account is to be found in the book of Raymond Pearl: *An introduction to medical biometry and statistics*, and in the paper of Gause and Alpatov ('31).

   2. Calculation of the equations of the struggle for existence.   We have deter-

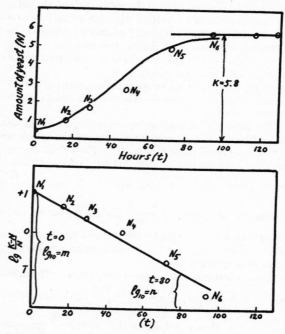

Fig. 41. Calculation of the logistic curve for the separate growth of the yeast, *Schizosaccharomyces kephir*, grown under anaerobic conditions (1931).

mined the maximal masses and the coefficients of multiplication: b_1, b_2, K_1, K_2, and following the instructions of Chapter IV we have calculated the coefficients of the struggle for existence α and β.   The curves corresponding to the system of the differential equations of the struggle for existence must now be found for a verification of the numerical values of α and β.

   We have a system of equations:

$$\frac{dN_1}{dt} = b_1 N_1 \frac{K_1 - (N_1 + \alpha N_2)}{K_1} \left.\right\}$$

$$\frac{dN_2}{dt} = b_2 N_2 \frac{K_2 - (N_2 + \beta N_1)}{K_2}$$

TABLE 8

Approximate integration of the system of equations of the struggle for existence with the aid of the method of Runge-Kutta (general form)

t	N_1	N_2	$\dfrac{dN_1}{dt}$	$\dfrac{dN_2}{dt}$	$k = \dfrac{dN_1}{dt}\cdot h$	$l = \dfrac{dN_2}{dt}\cdot h$	(k)	(l)
t	N_1	N_2	$b_1 N_1 \dfrac{K_1-(N_1+\alpha N_2)}{K_1}$	$b_2 N_2 \dfrac{K_2-(N_2+\beta N_1)}{K_2}$	k_1	l_1	$\frac{1}{2}(k_1+k_4)$	$\frac{1}{2}(l_1+l_4)$
$t+\dfrac{h}{2}$	$N_1+\dfrac{k_1}{2}$	$N_2+\dfrac{l_1}{2}$	$b_1\left(N_1+\dfrac{k_1}{2}\right)\dfrac{K_1-\left[\left(N_1+\dfrac{k_1}{2}\right)+\alpha\left(N_2+\dfrac{l_1}{2}\right)\right]}{K_1}$	$b_2\left(N_2+\dfrac{l_1}{2}\right)\dfrac{K_2-\left[\left(N_2+\dfrac{l_1}{2}\right)+\beta\left(N_1+\dfrac{k_1}{2}\right)\right]}{K_2}$	k_2	l_2	k_2+k_3 sum	l_2+l_3 sum
$t+\dfrac{h}{2}$	$N_1+\dfrac{k_2}{2}$	$N_2+\dfrac{l_2}{2}$	$b_1\left(N_1-\dfrac{k_2}{2}\right)\dfrac{K_1-\left[\left(N_1+\dfrac{k_2}{2}\right)+\alpha\left(N_2+\dfrac{l_2}{2}\right)\right]}{K_1}$	$b_2\left(N_2+\dfrac{l_2}{2}\right)\dfrac{K_2-\left[\left(N_2+\dfrac{l_2}{2}\right)+\beta\left(N_1+\dfrac{k_2}{2}\right)\right]}{K_2}$	k_3	l_3	$k=\frac{1}{3}$ of the sum	$l=\frac{1}{3}$ of the sum
$t+h$	N_1+k_3	N_2+l_3	$b_1(N_1+k_3)\dfrac{K_1-[(N_1+k_3)+\alpha(N_2+l_3)]}{K_1}$	$b_2(N_2+l_3)\dfrac{K_2-[(N_2+l_3)+\beta(N_1+k_3)]}{K_2}$	k_4	l_4		
$t+h$	N_1+k	N_2+l						

where the numerical values of the parameters are already known, and where it is necessary to calculate the values N_1 and N_2 corresponding to different moments of time. As this system of differential equations cannot be solved at present, we will use the method of approximate numerical integration of Runge-Kutta.

§1. *Ordinates at origin.* The system of differential equations does not include the special parameters characterizing the values N_1 and N_2 with $t = 0$ (corresponding to parameter a of the logistic curve). We know these values from the experiment, and they correspond to those quantities of the first and second species which are introduced into a mixed population at the very beginning. The same quantities are also introduced into the separate populations. Therefore, according to the logistic curves of separate growth we calculate the value N_1 with $t = 0$, and N_2 with $t = 0$. The quantities obtained are taken for the initial ordinates of the system of differential equations.

§2. *Calculations.* The meaning of the calculations is that knowing the value of N_1 and N_2 at $t = 0$, we give to time a certain increment h and for this new moment $(t + h)$ we calculate according to the system of the equations of competition the corresponding increments of the populations of the first and second species: $(N_1 + k)$ and $(N_2 + l)$. These calculations are made with the aid of the formulae of Runge-Kutta, which for our case take the form given in Table 8.

The initial values of the ordinates enable us to calculate $\dfrac{dN_1}{dt}$ and $\dfrac{dN_2}{dt}$. Multiplying them by h we obtain k_1, and l_1. Thereupon in the second line we calculate again $\dfrac{dN_1}{dt}$ and $\dfrac{dN_2}{dt}$ starting no longer from the initial ordinates but with $\left(N_1 + \dfrac{k_1}{2}\right)$ and $\left(N_2 + \dfrac{l_1}{2}\right)$. Multiplying the result by h we obtain k_2 and l_2, and continue to calculate until we obtain k_1, k_2, k_3, k_4 and l_1, l_2, l_3, l_4. We then pass on to the right side of the table. Here we calculate $\frac{1}{2}(k_1 + k_4)$ and then $k_2 + k_3$. Summing and taking $\frac{1}{3}$ of the sum we obtain the required increment of the first species equal to k. The increment of the second species is calculated in the same way.

The ordinates thus calculated $(N_1 + k)$ and $(N_2 + l)$ can in their turn be considered as already known initial values, and following the same plan we can calculate the new increments corresponding to the moment of time $(t + 2h)$. Continuing these calculations we will find the growth curves of the first and second species in a mixed population. Table 9 presents a numerical example of the calculation of such curves for *Saccharomyces* (No. 1) and *Schizosaccharomyces* (No. 2) in a mixed anaerobic culture according to the experiments of 1931.

§3. *Final values of N_1 and N_2.* The process of calculating will bring us to the moment when $\dfrac{dN_1}{dt}$ and $\dfrac{dN_2}{dt}$ will be near to zero, and we will approach the final values of N_1 and N_2 (the experiments described in Chapter IV were made in conditions of limited resources of energy, where on the disappearance of the unutilized opportunity growth simply ceased). If in the approximation to

TABLE 9

The numerical example of the approximate integration of the system of equations of the struggle for existence with the aid of the method of Runge-Kutta

$\alpha = 3.15$ $K_1 = 13.0$ units of volume $b_1 = 0.21827$ units per hour $h = 10$
$\beta = 0.439$ $K_2 = 5.8$ units of volume $b_2 = 0.06069$ units per hour   hours

t	N_1	N_2	$\dfrac{dN_1}{dt}$	$\dfrac{dN_2}{dt}$	k	l	(k)	(l)
0	*0.450*	*0.450*	0.0841	0.0242	0.841	0.242	2.114	0.275
5	0.871	0.571	0.1510	0.0289	1.510	0.289	*3.518*	*0.580*
5	1.205	0.594	0.2008	0.0291	2.008	0.291	5.632	0.855
10	2.458	0.741	0.3387	0.0309	3.387	0.309	1.877	0.285
10	*2.327*	*0.735*	0.3262	0.0311	3.262	0.311	3.406	0.255
15	3.958	0.890	0.4145	0.0295	4.145	0.295	*8.446*	*0.570*
15	4.399	0.882	0.4301	0.0275	4.301	0.275	11.852	0.825
20	6.628	1.010	0.3550	0.0199	3.550	0.199	3.951	0.275
20	*6.278*	*1.010*	0.3732	0.0215	3.732	0.215	1.911	0.143
25	8.144	1.117	0.1829	0.0129	1.829	0.129	*4.757*	*0.305*
25	7.192	1.074	0.2928	0.0176	2.928	0.176	6.668	0.448
30	9.206	1.186	0.0090	0.0071	0.090	0.071	2.223	0.149
30	*8.501*	*1.159*	0.1210	0.0110	1.210	0.110	0.605	0.080
35	9.106	1.214	0.0107	0.0075	0.107	0.075	*1.081*	*0.181*
35	8.554	1.196	0.0974	0.0106	0.974	0.106	1.686	0.261
40	9.475	1.265	0	0.0050	0	0.050	0.562	0.087
40	*9.063*	*1.246*						

the final value of N we pass beyond the asymptote K, and $\dfrac{dN}{dt}$ becomes negative, we will simply consider it as equal to zero. With the resources of energy maintained at a fixed level the calculations must apparently be continued according to the directions given in Chapter V. A detailed description of the method of Runge-Kutta is to be found in the book *Vorlesungen über numerische Rechnen* by Runge and König ('24).

BIBLIOGRAPHY[1]

1. ALECHIN, W. W.   1926   Was ist eine Pflanzengesellschaft?   Beih. Repert. spec. nov. regni veget. 37.
2. ALLEE, W. C.   1931   Animal aggregations.   Chicago.
3. ALLEE, W. C.   1932   A review of Chapman's "Animal ecology."   Ecology. 13, p. 405.
4. ALLEE, W. C.   1934   Recent studies in mass physiology.   Biol. Rev. 9, p. 1.
5. BAILEY, V. A.   1931   The interaction between hosts and parasites. Quart. Journ. Math. (Oxford Series) 2, p. 68.
6. BAILEY, V. A.   1933   On the interaction between several species of hosts and parasites.   Proc. Roy. Soc. A, 143, p. 75.
7. BARKER, H. A. AND TAYLOR, C. V.   1931   A study of the conditions of encystment of Colpoda cucullus.   Phys. Zool. 4, p. 620.
8. BEAUCHAMP, R. AND ULLYOTT, P.   1932   Competitive relationships between certain species of fresh-water triclads.   Journ. Ecol. 20, p. 200.
9. BEERS, C. D.   1927   The relation between hydrogen-ion concentration and encystment in Didinium nasutum.   J. Morph. 44, p. 21.
10. BEERS, C. D.   1933a   Diet in relation to depression and recovery in the ciliate, Didinium nasutum.   Arch. Protistenknd. 79, p. 101.
11. BEERS, C. D.   1933b   The relation of density of population to rate of reproduction in the ciliates, Didinium nasutum and Stylonychia pustulata.   Arch. Protistenknd. 80, p. 36.
12. BĚLAŘ, A.   1928   Untersuchung der Protozoen.   Peterfi Meth. wiss. Biol. 1, p. 735.
13. BIRSTEIN, J. A. UND VINOGRADOV, L. G.   1934   Die Süsswasserdecapoden der U. S. S. R. und ihre geographische Verbreitung.   Zool. Zeitschr. 13, p. 39 (in Russian).
14. BODENHEIMER, F. S.   1932   Über den Massenwechsel der Selachierbevölkerung im oberadriatischen Benthos.   Arch. Hydrob. 24, p. 667.
15. BOLTZMANN, L.   1905   Populäre Schriften.   Leipzig.
16. BRAUN-BLANQUET, J.   1928   Pflanzensoziologie.   Berlin.
17. BRUNELLI, G.   1929   Equilibri biologici e preteso spolamento del mare. Riv. Biol. 11, p. 779–785.
18. CALCINS, G. N.   1915   Didinium nasutum.   I. The life history.   Journ. Exp. Zool. 19, p. 225.
19. CALCINS, G. N.   1933   The biology of Protozoa.   Second Edition.
20. CHAPMAN, R. N.   1928   The quantitative analysis of environmental factors.   Ecol. 9, p. 111.
21. CHAPMAN, R. N.   1931   Animal ecology.   New York.

---

[1] This list of literature includes mainly references cited in the text.

154

22. CHINCHIN, A. J. 1929 The principles of physical statistics. Progr. Phys. Sci. 9, p. 141 (in Russian).

23. CLEMENTS, F. E. AND WEAVER, J. E. 1924 Experimental vegetation. Carn. Inst. Washington.

24. CLEMENTS, F. E., WEAVER, J. E. AND HANSON, H. C. 1929 Plant competition. Carn. Inst. Washington.

25. COCKERELL, T. D. A. 1934 "Mimicry" among insects. Nature. 133, p. 329.

26. CUTLER, D. W. 1923 Ann. App. Biol. 10, p. 137.

27. D'ANCONA, U. 1926 Dell 'influenza della stasi peschereccia del periodo 1914–1918 sul patrimonio ittico dell' Alto Adriatico. R. Comitato Talass. Italiano. Mem. 126.

28. D'ANCONA, U. 1927 Intorno alle associazioni biologiche e a un saggio di teoria matematica sulle stesse con particolar riguardo all'idrobiologia. Int. Rev. Ges. Hydr. Hydrograph. 17, p. 189.

29. DARBY, H. H. 1929 The effect of the hydrogen ion concentration on the sequence of protozoan forms. Arch. Protistenkunde. 65, p. 1.

30. DARWIN, CH. 1859 The origin of species by means of natural selection (cited from the edition of 1899, London: John Murray).

31. DECANDOLLE, A. P. 1820 Essai élémentaire de géographie botanique.

32. DEWAR, D. 1926 Indian bird life: The struggle for existence of birds in India. Pp. 292.

33. DUFFIELD, J. E. 1933 Journ. Anim. Ecol. 2, p. 184.

34. DU-RIETZ, G. E. 1930 Vegetationsforschung auf soziationsanalytischen Grundlage. Abderh. Handb. d. biol. Arbeitsmeth. Abt. XI, Teil 5, H. 2.

35. EDDY, S. 1928 Succession of Protozoa in cultures under controlled conditions. Transact. Am. Micr. Soc. 47, p. 283.

36. ELTON, C. 1927 Animal Ecology. London.

37. EMBODY, G. C. 1933 Cited from a letter.

38. FISHER, R. A. 1930 The genetical theory of natural selection. Oxford.

39. FORD, E. B. 1930 Fluctuation in numbers, and its influence on variation in Melitaea aurinia. Trans. Ent. Soc. Lond. 78, p. 345.

40. FORMOSOV, A. N. 1934 Cited from manuscript.

41. GAUSE, G. F. UND ALPATOV, W. W. 1931 Die logistische Kurve von Verhulst-Pearl und ihre Anwendung im Gebiet der quantitativen Biologie. Biol. Zentralbl. 51, p. 1.

42. GAUSE, G. F. 1932a Ecology of populations. Quart. Rev. Biol. 7, p. 27.

43. GAUSE, G. F. 1932b Experimental studies on the struggle for existence. I. Mixed population of two species of yeast. Journ. Exp. Biol. (British) 9, p. 389.

44. GAUSE, G. F. 1933 Über die Kinetik der Akkumulation der organischen Substanz in aus zwei Hefearten bestehenden Kulturen. Bioch. Zeitschr. 266, p. 352.

45. GAUSE, G. F. 1934a. Experimental analysis of Vito Volterra's mathematical theory of the struggle for existence. Science. 79, p. 16.

46. GAUSE, G. F. 1934b Über die Konkurrenz zwischen zwei Arten. Zool. Anz. 105, p. 219.
47. GAUSE, G. F. 1934c Über einige quantitative Beziehungen in der Insekten-Epidemiologie. Zeitschr. ang. Entom. 20, p. 619.
48. GAUSE, G. F. 1934d Experimentelle Untersuchungen über Konkurrenz zwischen Paramecium caudatum und Paramecium aurelia. Arch. Protistenkunde (in press), also in Revue Zool. Russe, 13, p. 1 (1934).
49. GAUSE, G. F. 1935a Untersuchungen über den Kampf ums Dasein bei Protisten. Biol. Zentralbl. 55 (in press) also in Revue Zool. Russe, 13, p. 18 (1934).
50. GAUSE, G. F. 1935b Experimental demonstration of Volterra's periodic oscillations in the numbers of animals. Journ. Exp. Biol. (British), 12 (in press).
51. GAUSE, G. F., NASTUKOVA, O. K. AND ALPATOV, W. W. 1935 The influence of biologically conditioned media on the growth of a mixed population of Paramecium caudatum and Paramecium aurelia. Journ. Anim. Ecol. (in press).
52. GAUSE, G. F. AND WITT, A. A. 1935 On the periodic fluctuations in the numbers of animals: A mathematical theory of the relaxation interaction between predators and prey, and its application to a population of Protozoa (in press).
53. GOLDMAN, E. A. 1930 The coyote-archpredator. Sympos. on predat. anim. control. J. Mammal. 11, p. 325.
54. GRAY, J. 1929 The kinetics of growth. Brit. Journ. Exp. Biol. 6, p. 248.
55. GRAY, J. 1931 A Text-book of experimental cytology. London.
56. GREENWOOD, M. 1932 Epidemiology. Historical and experimental. Baltimore.
57. HALDANE, J. B. S. 1924 Trans. Camb. Phil. Soc. 23, p. 19.
58. HALDANE, J. B. S. 1931 Proc. Camb. Phil. Soc. 27, p. 131.
59. HALDANE, J. B. S. 1932 The causes of evolution. London.
60. HARGITT, G. T. AND FRAY, W. 1927 The growth of Paramecium in pure cultures of bacteria. J. Exp. Zool. 22, p. 421.
61. HARTMANN, M. 1927 Allgemeine Biologie. Jena.
62. HILL, A. V. 1933 Wave transmission as the basis of nerve activity. Scient. Month. p. 316.
63. H. T. H. P. 1931 A review of Prof. Volterra's book "Leçons sur la théorie mathématique de la lutte pour la vie." Nature. 128, p. 963.
64. JENNINGS, H. S. 1908 Heredity, variation and evolution in Protozoa. III. Proc. Am. Phil. Soc. 47, p. 393.
65. JENNINGS, H. S. 1915 Behavior of the lower organisms. New York.
66. JENNINGS, H. S. 1929 Genetics of the Protozoa. Bibliogr. Genet. 5, p. 105.
67. JENNINGS, H. S. 1933 Genetics of the Protozoa, in relation to some of the greater problems of genetics. Japan. J. Genet. 8, p. 65.
68. JENSEN, A. J. C. 1933 Periodic fluctuations in the size of various stocks of fish and their causes. Medd. Komm. Havundersogelser. Fisk. 9, p. 5.

69. JOHNSON, W. H.   1933   Effects of population density of the rate of repro-
    duction in Oxytricha.   Phys. Zool. 6, p. 22.
70. JOLLOS, V.   1921   Experimentelle Protistenstudien.   I. Arch. Protisten-
    kunde. 43, p. 1.
71. KALABUCHOV, N. J. UND RAEWSKI, W.   1933   Zur Methodik des Stu-
    diums einiger Fragen der Ökologie mäuseartiger Nager.   Rev. Microb.
    Epid. et Parasit. 12, p. 47 (in Russian).
72. KALMUS, H.   1931   Paramecium. Das Pantoffeltierchen. Jena.
73. KASHKAROV, D. N.   1928   Ecological survey of the environments of the
    lakes Beely-Kul, etc., Turkestan.   Acta Univ. Asiae Mediae, Zool.,
    2 (in Russian).
74. KERMACK, W. O. AND MCKENDRICK, A. J.   1927   A contribution to the
    mathematical theory of epidemics.   Proc. Roy. Soc. A. 115, p. 700.
75. KESSLER, J. TH.   1875   Russian crayfishes. Transact. Russ. Ent. Soc.
    8, p. 228 (in Russian).
76. KLEM, A.   1933   On the growth of populations of yeast.   Hvalradets
    Skrifter (Scientific Results of Marine Biological Research) Oslo. 7,
    p. 55.
77. LASAREFF, P. P.   1923   La théorie ionique de l'excitation.   Moscou (in
    Russian).
78. L'HÉRITIER, PH. ET TEISSIER, G.   1933   Etude d'une population de
    Drosophiles en equilibre.   C. R. Ac. Sci. 197, p. 1765.
79. L'HÉRITIER, PH.   1934   Etude démorgaphique comparée de quatre
    lignées de Drosophila melanogaster.   C. R. Ac. Sci. 198, p. 770.
80. LOTKA, A. J.   1910   Contribution to the theory of periodic reactions.   J.
    Phys. Chem. 14, p. 271.
81. LOTKA, A. J.   1920a   Undamped oscillations derived from the law of mass
    action.   Journ. Americ. Chem. Soc. 52, p. 1595.
82. LOTKA, A. J.   1920b   Analytical note on certain rhythmic relations in
    organic systems.   Proc. Natl. Acad. 6, p. 410.
83. LOTKA, A. J.   1923a   Contribution to quantitative parasitology.   Journ.
    Wash. Acad. Sc. 13, p. 152.
84. LOTKA, A. J.   1923b   Contribution to the analysis of malaria epidemi-
    ology.   Amer. Journ. Hyg. 3, January Supplement, p. 1.
85. LOTKA, A. J.   1925   Elements of physical biology. Baltimore.
86. LOTKA, A. J.   1932a   Contribution to the mathematical theory of cap-
    ture.   I. Conditions for capture.   Proc. Nat. Ac. Sci. 18, p. 172.
87. LOTKA, A. J.   1932b   The growth of mixed populations: Two species com-
    peting for a common food supply.   Journ. Wash. Ac. Sci. 22, p. 461.
88. LUDWIG, W.   1933   Der Effekt der Selektion bei Mutationen geringen
    Selektionswerts.   Biol. Zbl. 53, p. 364.
89. MARCHI, C.   1928   Osservazioni sulla statistica della pesca di mare e di
    stagno a Cagliari dal 1912 al 1927.   Scritti biol. 4, p. 37.
90. MARCHI, C.   1929   Verifica pratica delle leggi teoriche di Vito Volterra
    sulle fluttuazioni del numero di individui in specie animali conviventi.
    R. Com. Talassogr. Ital. Mem. 154.

91. MAST, S. O. 1909 The reactions of Didinium nasutum with special reference to the feeding habits and the function of trichocysts. Biol. Bull. 16, p. 91.

92. MONTGOMERY, E. G. 1912 Competition in cereals. Nebr. Agr. Exp. Sta. Bull. 127, p. 3.

93. MYERS, E. C. 1927 Relation of density of population and certain other factors to survival and reproduction in different biotypes of Paramecium caudatum. J. Exp. Zool., 49, p. 1.

94. NÄGELI, C. 1874 Verdrängung der Pflanzenformen durch ihre Mitbewerber. Sitzb. Akad. Wiss. München 11, p. 109.

95. NICHOLSON, A. J. 1933 The balance of animal populations. Journ. Anim. Ecol. 2, p. 132.

96. OEHLER, R. 1920 Gereinigte Ciliatenzucht. Arch. Protistenkunde. 41, p. 34.

97. OSTERHOUT, W. J. V. 1906 On the importance of physiological balanced solutions for plants. Bot. Gaz. 42, p. 127.

98. PAVLOV, J. P. 1923 Twenty years of objective study of the higher nervous activity (behaviour) of animals. Conditional reflexes. Leningrad (in Russian).

99. PAYNE, N. M. 1933 The differential effect of environmental factors upon Microbracon hebetor and its host Ephestia Kühniella. Biol. Bull. 65, p. 187. Part II. Ecol. Monogr. (1934).

100. PEARL, R. AND REED, L. J. 1920 On the rate of growth of the population of the United States since 1790 and its mathematical representation. Proc. Nat. Acad. Sci. 6, p. 275.

101. PEARL, R. 1928 The rate of living. New York.

102. PEARL, R. 1930 Introduction to medical biometry and statistics. 2nd edition. Philadelphia.

103. PHILPOT, C. H. 1928 Growth of Paramecium in pure cultures of pathogenic bacteria and in the presence of soluble products of such bacteria. J. Morph. 46, p. 85.

104. PLATE, L. 1913 Selektionsprinzip und Probleme der Artbildung. Vierte Auflage. Leipzig und Berlin.

105. REUKAUF, E. 1930 Zur Biologie von Didinium nasutum. Zeitschr. Vergl. Physiol. 11, p. 689.

106. RICHARDS, O. W. 1928a Potentially unlimited multiplication of yeast with constant environment, and the limiting of growth by changing environment. Journ. Gen. Physiol. 11, p. 525.

107. RICHARDS, O. W. 1928b Changes in sizes of yeast cells during multiplication. Botan. Gaz. 86, p. 93.

108. RICHARDS, O. W. 1928c The growth of the yeast Saccharomyces cerevisiae. I. Ann. Bot. 42, p. 271.

109. RICHARDS, O. W. AND ROOPE, P. L. 1930 A tangent meter for graphical differentiation. Science. 71, p. 290.

110. RICHARDS, O. W. 1932 The second cycle and subsequent growth of a population of yeast. Arch. Protistenkunde. 78, p. 263.

111. ROSS, R. 1908 Report on the prevention of malaria in Mauritius.

112. Ross, R. 1911 The prevention of malaria. London.
113. Runge, C. und König, H. 1924 Vorlesungen über numerische Rechnen. Berlin.
114. Russell, E. J. 1927 Soil conditions and plant growth. London.
115. Sandon, H. 1932 The food of Protozoa. Cairo.
116. Severtzov, N. A. 1855 Periodic phenomena in the life of animals in Voronesh district. Moscow (in Russian).
117. Severtzov, S. 1933 Zum Problem der Dynamik der Herde bei den Wirbelthieren. Bull. Ac. Sci. U. S. S. R. p. 1005.
118. Shelford, V. E. 1931 Some concepts of bioecology. Ecol. 12, p. 455.
119. Skadovsky, S. N. 1915 The alteration of the reaction of environment in cultures of Protozoa. Transact. Univ. Shaniavsky. Biol. Lab. 1, p. 157 (in Russian).
120. Slator, A. 1913 Rate of fermentation by growing yeast cells. Bioch. Journ. 7, p. 197.
121. Smirnov, E. und Wladimirow, M. 1934 Studien über die Vermehrungsfähigkeit der Pteromalide Mormoniella vitripennis Wlk. (in press).
122. Stanley, J. 1932 A mathematical theory of the growth of populations of the flour beetle, Tribolium confusum. Canad. Journ. Res. 6, p. 632.
123. Sukatschew, W. N. 1927 Einige experimentelle Untersuchungen über den Kampf ums Dasein zwischen Biotypen derselben Art. Zeitsch. Ind. Abst. Vererb. 47, p. 54.
124. Sukatschew, W. N. 1928 Plant communities. Moscow (in Russian).
125. Tansley, A. G. 1917 On competition between Galium saxalite L. and G. silvestre Poll. on different types of soil. J. Ecol. 5, p. 173.
126. Thomson, G. M. 1922 The naturalisation of animals and plants in New-Zealand. Cambridge.
127. Timoféeff-Ressovsky, N. W. 1933 Über die relative Vitalität von Drosophila melanogaster und Drosophila funebris unter verschiedenen Zuchtbedingungen, in Zusammenhang mit den Verbreitungsarealen dieser Arten. Arch. Naturgesch. 2, p. 285.
128. Topley, W. W. C. 1926 Three Milroy Lectures on experimental epidemiology. The Lancet, p. 477; 531; 645.
129. Verhulst, P. F. 1839 Notice sur la loi que la population suit dans son accroissement. Corr. math. et phys. publ. par A. Quetelet. 10, p. 113.
130. Vernadsky, V. J. 1926 The Biosphere. Leningrad (in Russian).
131. Volterra, V. 1926 Variazioni e fluttuazioni del numero d'individui in specie animali conviventi. Mem. R. Accad. Naz. dei Lincei. Ser. VI, vol. 2.
132. Volterra, V. 1931 Leçons sur la théorie mathématique de la lutte pour la vie. Paris.
133. Warming, E. 1895 Lehrbuch der ökologischen Pflanzengeographie.
134. Winsor, C. P. 1932 The Gompertz curve as a growth curve. Proc. Nat. Acad. Sci. 18, p. 1.

135. Wood, F. E.   1910   A study of the mammals of Champaign Country, Illinois.   Bull. Ill. Nat. Hist. Surv. 8, p. 501.
136. Woodruff, L. L.   1911   The effect of excretion products of Paramecium on its rate of reproduction.   Journ. Exp. Zool. 10, p. 551.
137. Woodruff, L. L.   1912   Observations on the origin and sequence of the protozoan fauna of hay infusions.   Journ. Exp. Zool. 12, p. 205.
138. Woodruff, L. L.   1914   The effect of excretion products of Infusoria on the same and on different species, with special reference to the protozoan sequence in infusions.   Journ. Exp. Zool. 14, p. 575.

INDEX

Adaptation, in plant communities, 17.

Adaptation, in protozoan populations, 132 et seq., 140.

Alechin, W. W., 17, 18.

Allee, W. C., vi, 11.

Alpatov, W. W., 105, 111, 150.

Bacillus pyocyaneus, 97, 105, 118, 125.

Bacillus subtilis, 92.

Baer, von, 8.

Bailey, V. A., 25, 57.

Barker, H. A., 96.

Beauchamp, R., 17.

Beers, C. D., 154.

Belar, A., 96.

Biotic experiments, 13.

Birstein, J. A., 22.

Bodenheimer, F. S., 25.

Boltzmann, L., 90, 91.

Braun-Blanquet, J., 18.

Brunelli, G., 25.

Buffon, 8.

Bumpus, v.

Calcins, G. N., 114.

Chance, its rôle in the struggle for existence, 18, 19, 124, 125.

Chapman, R. N., 33.

Chinchin, A. J., 125.

Classification of the struggle for existence, 3.

Clements, F. E., 4, 5, 15.

Cockerell, T. D. A., 128.

Coefficient of selection, 111.

Coefficients of the struggle for existence, their calculation, 80.

Competition in field conditions, its complexity, 3, 12, 13, 111.

Cutler, D. W., 138, 139.

Cyprinus albus, 21.

D'Ancona, U., 25.

Darby, H. H., 104.

Darwin, Ch., v, 1, 2, 8, 14, 40.

De Candolle, A. P., 17.

Dewar, D., 155.

Didinium nasutum, 114 et seq., 132 et seq.

Direct struggle for existence, definition, 3.

Drosophila funebris, 17.

Drosophila melanogaster, 17.

Duffield, J. E., 128.

Du-Rietz, G. E., 19.

Eddy, S., 91.

Ehrenberg, 8.

Elementary process of the struggle for existence, definition, 2.

Elton, Ch., 12, 19.

Embody, G. C., 21.

Empirical equations, their meaning, 59, 60.

Ephestia, 25.

Euphorbia, 18.

Equation of competition, 44 et seq.

Experimental study of the struggle for existence, methods of, 6, 62 et seq., 92, 96 et seq.

Fisher, R. A., 155.

Ford, E. B., 111.

Formosov, A. N., 19.

Fray, W., 96.

Gause, G. F., 44, 50, 62, 68, 72, 77 et seq., 94, 101 et seq., 111, 118 et seq., 133 et seq., 137 et seq., 150.

Gibbs, 10.

Goldman, E. A., 22.

Gray, J., 43.

Greenwood, M., 114.

161

A CATALOGUE OF SELECTED DOVER BOOKS
IN ALL FIELDS OF INTEREST

A CATALOGUE OF SELECTED DOVER BOOKS
IN ALL FIELDS OF INTEREST

AMERICA'S OLD MASTERS, James T. Flexner. Four men emerged unexpectedly from provincial 18th century America to leadership in European art: Benjamin West, J. S. Copley, C. R. Peale, Gilbert Stuart. Brilliant coverage of lives and contributions. Revised, 1967 edition. 69 plates. 365pp. of text.
21806-6 Paperbound $2.75

FIRST FLOWERS OF OUR WILDERNESS: AMERICAN PAINTING, THE COLONIAL PERIOD, James T. Flexner. Painters, and regional painting traditions from earliest Colonial times up to the emergence of Copley, West and Peale Sr., Foster, Gustavus Hesselius, Feke, John Smibert and many anonymous painters in the primitive manner. Engaging presentation, with 162 illustrations. xxii + 368pp.
22180-6 Paperbound $3.50

THE LIGHT OF DISTANT SKIES: AMERICAN PAINTING, 1760-1835, James T. Flexner. The great generation of early American painters goes to Europe to learn and to teach: West, Copley, Gilbert Stuart and others. Allston, Trumbull, Morse; also contemporary American painters—primitives, derivatives, academics—who remained in America. 102 illustrations. xiii + 306pp.
22179-2 Paperbound $3.00

A HISTORY OF THE RISE AND PROGRESS OF THE ARTS OF DESIGN IN THE UNITED STATES, William Dunlap. Much the richest mine of information on early American painters, sculptors, architects, engravers, miniaturists, etc. The only source of information for scores of artists, the major primary source for many others. Unabridged reprint of rare original 1834 edition, with new introduction by James T. Flexner, and 394 new illustrations. Edited by Rita Weiss. 6⅝ x 9⅝.
21695-0, 21696-9, 21697-7 Three volumes, Paperbound $13.50

EPOCHS OF CHINESE AND JAPANESE ART, Ernest F. Fenollosa. From primitive Chinese art to the 20th century, thorough history, explanation of every important art period and form, including Japanese woodcuts; main stress on China and Japan, but Tibet, Korea also included. Still unexcelled for its detailed, rich coverage of cultural background, aesthetic elements, diffusion studies, particularly of the historical period. 2nd, 1913 edition. 242 illustrations. lii + 439pp. of text.
20364-6, 20365-4 Two volumes, Paperbound $5.00

THE GENTLE ART OF MAKING ENEMIES, James A. M. Whistler. Greatest wit of his day deflates Oscar Wilde, Ruskin, Swinburne; strikes back at inane critics, exhibitions, art journalism; aesthetics of impressionist revolution in most striking form. Highly readable classic by great painter. Reproduction of edition designed by Whistler. Introduction by Alfred Werner. xxxvi + 334pp.
21875-9 Paperbound $2.25

VISUAL ILLUSIONS: THEIR CAUSES, CHARACTERISTICS, AND APPLICATIONS, Matthew Luckiesh. Thorough description and discussion of optical illusion, geometric and perspective, particularly; size and shape distortions, illusions of color, of motion; natural illusions; use of illusion in art and magic, industry, etc. Most useful today with op art, also for classical art. Scores of effects illustrated. Introduction by William H. Ittleson. 100 illustrations. xxi + 252pp.

21530-X Paperbound $1.50

A HANDBOOK OF ANATOMY FOR ART STUDENTS, Arthur Thomson. Thorough, virtually exhaustive coverage of skeletal structure, musculature, etc. Full text, supplemented by anatomical diagrams and drawings and by photographs of undraped figures. Unique in its comparison of male and female forms, pointing out differences of contour, texture, form. 211 figures, 40 drawings, 86 photographs. xx + 459pp. 5⅜ x 8⅜.

21163-0 Paperbound $3.00

150 MASTERPIECES OF DRAWING, Selected by Anthony Toney. Full page reproductions of drawings from the early 16th to the end of the 18th century, all beautifully reproduced: Rembrandt, Michelangelo, Dürer, Fragonard, Urs, Graf, Wouwerman, many others. First-rate browsing book, model book for artists. xviii + 150pp. 8⅜ x 11¼.

21032-4 Paperbound $2.00

THE LATER WORK OF AUBREY BEARDSLEY, Aubrey Beardsley. Exotic, erotic, ironic masterpieces in full maturity: Comedy Ballet, Venus and Tannhauser, Pierrot, Lysistrata, Rape of the Lock, Savoy material, Ali Baba, Volpone, etc. This material revolutionized the art world, and is still powerful, fresh, brilliant. With *The Early Work*, all Beardsley's finest work. 174 plates, 2 in color. xiv + 176pp. 8⅛ x 11.

21817-1 Paperbound $2.75

DRAWINGS OF REMBRANDT, Rembrandt van Rijn. Complete reproduction of fabulously rare edition by Lippmann and Hofstede de Groot, completely reedited, updated, improved by Prof. Seymour Slive, Fogg Museum. Portraits, Biblical sketches, landscapes, Oriental types, nudes, episodes from classical mythology—All Rembrandt's fertile genius. Also selection of drawings by his pupils and followers. "Stunning volumes," *Saturday Review*. 550 illustrations. lxxviii + 552pp. 9⅛ x 12¼.

21485-0, 21486-9 Two volumes, Paperbound $6.50

THE DISASTERS OF WAR, Francisco Goya. One of the masterpieces of Western civilization—83 etchings that record Goya's shattering, bitter reaction to the Napoleonic war that swept through Spain after the insurrection of 1808 and to war in general. Reprint of the first edition, with three additional plates from Boston's Museum of Fine Arts. All plates facsimile size. Introduction by Philip Hofer, Fogg Museum. v + 97pp. 9⅜ x 8¼.

21872-4 Paperbound $1.75

GRAPHIC WORKS OF ODILON REDON. Largest collection of Redon's graphic works ever assembled: 172 lithographs, 28 etchings and engravings, 9 drawings. These include some of his most famous works. All the plates from *Odilon Redon: oeuvre graphique complet*, plus additional plates. New introduction and caption translations by Alfred Werner. 209 illustrations. xxvii + 209pp. 9⅛ x 12¼.

21966-8 Paperbound $4.00

DESIGN BY ACCIDENT; A BOOK OF "ACCIDENTAL EFFECTS" FOR ARTISTS AND DESIGNERS, James F. O'Brien. Create your own unique, striking, imaginative effects by "controlled accident" interaction of materials: paints and lacquers, oil and water based paints, splatter, crackling materials, shatter, similar items. Everything you do will be different; first book on this limitless art, so useful to both fine artist and commercial artist. Full instructions. 192 plates showing "accidents," 8 in color. viii + 215pp. 8⅜ x 11¼. 21942-9 Paperbound $3.50

THE BOOK OF SIGNS, Rudolf Koch. Famed German type designer draws 493 beautiful symbols: religious, mystical, alchemical, imperial, property marks, ١nes, etc. Remarkable fusion of traditional and modern. Good for suggestions of timelessness, smartness, modernity. Text. vi + 104pp. 6⅛ x 9¼. 20162-7 Paperbound $1.25

HISTORY OF INDIAN AND INDONESIAN ART, Ananda K. Coomaraswamy. An unabridged republication of one of the finest books by a great scholar in Eastern art. Rich in descriptive material, history, social backgrounds; Sunga reliefs, Rajput paintings, Gupta temples, Burmese frescoes, textiles, jewelry, sculpture, etc. 400 photos. viii + 423pp. 6⅜ x 9¾. 21436-2 Paperbound $3.50

PRIMITIVE ART, Franz Boas. America's foremost anthropologist surveys textiles, ceramics, woodcarving, basketry, metalwork, etc.; patterns, technology, creation of symbols, style origins. All areas of world, but very full on Northwest Coast Indians. More than 350 illustrations of baskets, boxes, totem poles, weapons, etc. 378 pp. 20025-6 Paperbound $2.50

THE GENTLEMAN AND CABINET MAKER'S DIRECTOR, Thomas Chippendale. Full reprint (third edition, 1762) of most influential furniture book of all time, by master cabinetmaker. 200 plates, illustrating chairs, sofas, mirrors, tables, cabinets, plus 24 photographs of surviving pieces. Biographical introduction by N. Bienenstock. vi + 249pp. 9⅞ x 12¾. 21601-2 Paperbound $3.50

AMERICAN ANTIQUE FURNITURE, Edgar G. Miller, Jr. The basic coverage of all American furniture before 1840. Individual chapters cover type of furniture—clocks, tables, sideboards, etc.—chronologically, with inexhaustible wealth of data. More than 2100 photographs, all identified, commented on. Essential to all early American collectors. Introduction by H. E. Keyes. vi + 1106pp. 7⅞ x 10¾. 21599-7, 21600-4 Two volumes, Paperbound $7.50

PENNSYLVANIA DUTCH AMERICAN FOLK ART, Henry J. Kauffman. 279 photos, 28 drawings of tulipware, Fraktur script, painted tinware, toys, flowered furniture, quilts, samplers, hex signs, house interiors, etc. Full descriptive text. Excellent for tourist, rewarding for designer, collector. Map. 146pp. 7⅞ x 10¾. 21205-X Paperbound $2.00

EARLY NEW ENGLAND GRAVESTONE RUBBINGS, Edmund V. Gillon, Jr. 43 photographs, 226 carefully reproduced rubbings show heavily symbolic, sometimes macabre early gravestones, up to early 19th century. Remarkable early American primitive art, occasionally strikingly beautiful; always powerful. Text. xxvi + 207pp. 8⅜ x 11¼. 21380-3 Paperbound $3.00

ALPHABETS AND ORNAMENTS, Ernst Lehner. Well-known pictorial source for decorative alphabets, script examples, cartouches, frames, decorative title pages, calligraphic initials, borders, similar material. 14th to 19th century, mostly European. Useful in almost any graphic arts designing, varied styles. 750 illustrations. 256pp. 7 x 10.   21905-4 Paperbound $3.50

PAINTING: A CREATIVE APPROACH, Norman Colquhoun. For the beginner simple guide provides an instructive approach to painting: major stumbling blocks for beginner; overcoming them, technical points; paints and pigments; oil painting; watercolor and other media and color. New section on "plastic" paints. Glossary. Formerly *Paint Your Own Pictures*. 221pp.   22000-1 Paperbound $1.75

THE ENJOYMENT AND USE OF COLOR, Walter Sargent. Explanation of the relations between colors themselves and between colors in nature and art, including hundreds of little-known facts about color values, intensities, effects of high and low illumination, complementary colors. Many practical hints for painters, references to great masters. 7 color plates, 29 illustrations. x + 274pp.

20944-X Paperbound $2.50

THE NOTEBOOKS OF LEONARDO DA VINCI, compiled and edited by Jean Paul Richter. 1566 extracts from original manuscripts reveal the full range of Leonardo's versatile genius: all his writings on painting, sculpture, architecture, anatomy, astronomy, geography, topography, physiology, mining, music, etc., in both Italian and English, with 186 plates of manuscript pages and more than 500 additional drawings. Includes studies for the Last Supper, the lost Sforza monument, and other works. Total of xlvii + 866pp. $7\frac{7}{8}$ x $10\frac{3}{4}$.

22572-0, 22573-9 Two volumes, Paperbound $10.00

MONTGOMERY WARD CATALOGUE OF 1895. Tea gowns, yards of flannel and pillow-case lace, stereoscopes, books of gospel hymns, the New Improved Singer Sewing Machine, side saddles, milk skimmers, straight-edged razors, high-button shoes, spittoons, and on and on . . . listing some 25,000 items, practically all illustrated. Essential to the shoppers of the 1890's, it is our truest record of the spirit of the period. Unaltered reprint of Issue No. 57, Spring and Summer 1895. Introduction by Boris Emmet. Innumerable illustrations. xiii + 624pp. $8\frac{1}{2}$ x $11\frac{5}{8}$.

22377-9 Paperbound $6.95

THE CRYSTAL PALACE EXHIBITION ILLUSTRATED CATALOGUE (LONDON, 1851). One of the wonders of the modern world—the Crystal Palace Exhibition in which all the nations of the civilized world exhibited their achievements in the arts and sciences—presented in an equally important illustrated catalogue. More than 1700 items pictured with accompanying text—ceramics, textiles, cast-iron work, carpets, pianos, sleds, razors, wall-papers, billiard tables, beehives, silverware and hundreds of other artifacts—represent the focal point of Victorian culture in the Western World. Probably the largest collection of Victorian decorative art ever assembled— indispensable for antiquarians and designers. Unabridged republication of the Art-Journal Catalogue of the Great Exhibition of 1851, with all terminal essays. New introduction by John Gloag, F.S.A. xxxiv + 426pp. 9 x 12.

22503-8 Paperbound $4.50

A HISTORY OF COSTUME, Carl Köhler. Definitive history, based on surviving pieces of clothing primarily, and paintings, statues, etc. secondarily. Highly readable text, supplemented by 594 illustrations of costumes of the ancient Mediterranean peoples, Greece and Rome, the Teutonic prehistoric period; costumes of the Middle Ages, Renaissance, Baroque, 18th and 19th centuries. Clear, measured patterns are provided for many clothing articles. Approach is practical throughout. Enlarged by Emma von Sichart. 464pp. 21030-8 Paperbound $3.00

ORIENTAL RUGS, ANTIQUE AND MODERN, Walter A. Hawley. A complete and authoritative treatise on the Oriental rug—where they are made, by whom and how, designs and symbols, characteristics in detail of the six major groups, how to distinguish them and how to buy them. Detailed technical data is provided on periods, weaves, warps, wefts, textures, sides, ends and knots, although no technical background is required for an understanding. 11 color plates, 80 halftones, 4 maps. vi + 320pp. $6\frac{1}{8}$ x $9\frac{1}{8}$. 22366-3 Paperbound $5.00

TEN BOOKS ON ARCHITECTURE, Vitruvius. By any standards the most important book on architecture ever written. Early Roman discussion of aesthetics of building, construction methods, orders, sites, and every other aspect of architecture has inspired, instructed architecture for about 2,000 years. Stands behind Palladio, Michelangelo, Bramante, Wren, countless others. Definitive Morris H. Morgan translation. 68 illustrations. xii + 331pp. 20645-9 Paperbound $2.50

THE FOUR BOOKS OF ARCHITECTURE, Andrea Palladio. Translated into every major Western European language in the two centuries following its publication in 1570, this has been one of the most influential books in the history of architecture. Complete reprint of the 1738 Isaac Ware edition. New introduction by Adolf Placzek, Columbia Univ. 216 plates. xxii + 110pp. of text. $9\frac{1}{2}$ x $12\frac{3}{4}$. 21308-0 Clothbound $10.00

STICKS AND STONES: A STUDY OF AMERICAN ARCHITECTURE AND CIVILIZATION, Lewis Mumford.One of the great classics of American cultural history. American architecture from the medieval-inspired earliest forms to the early 20th century; evolution of structure and style, and reciprocal influences on environment. 21 photographic illustrations. 238pp. 20202-X Paperbound $2.00

THE AMERICAN BUILDER'S COMPANION, Asher Benjamin. The most widely used early 19th century architectural style and source book, for colonial up into Greek Revival periods. Extensive development of geometry of carpentering, construction of sashes, frames, doors, stairs; plans and elevations of domestic and other buildings. Hundreds of thousands of houses were built according to this book, now invaluable to historians, architects, restorers, etc. 1827 edition. 59 plates. 114pp. $7\frac{7}{8}$ x $10\frac{3}{4}$. 22236-5 Paperbound $3.00

DUTCH HOUSES IN THE HUDSON VALLEY BEFORE 1776, Helen Wilkinson Reynolds. The standard survey of the Dutch colonial house and outbuildings, with constructional features, decoration, and local history associated with individual homesteads. Introduction by Franklin D. Roosevelt. Map. 150 illustrations. 469pp. $6\frac{5}{8}$ x $9\frac{1}{4}$. 21469-9 Paperbound $3.50

THE ARCHITECTURE OF COUNTRY HOUSES, Andrew J. Downing. Together with Vaux's *Villas and Cottages* this is the basic book for Hudson River Gothic architecture of the middle Victorian period. Full, sound discussions of general aspects of housing, architecture, style, decoration, furnishing, together with scores of detailed house plans, illustrations of specific buildings, accompanied by full text. Perhaps the most influential single American architectural book. 1850 edition. Introduction by J. Stewart Johnson. 321 figures, 34 architectural designs. xvi + 560pp.
22003-6 Paperbound $3.50

LOST EXAMPLES OF COLONIAL ARCHITECTURE, John Mead Howells. Full-page photographs of buildings that have disappeared or been so altered as to be denatured, including many designed by major early American architects. 245 plates. xvii + 248pp. 7⅞ x 10¾.
21143-6 Paperbound $3.00

DOMESTIC ARCHITECTURE OF THE AMERICAN COLONIES AND OF THE EARLY REPUBLIC, Fiske Kimball. Foremost architect and restorer of Williamsburg and Monticello covers nearly 200 homes between 1620-1825. Architectural details, construction, style features, special fixtures, floor plans, etc. Generally considered finest work in its area. 219 illustrations of houses, doorways, windows, capital mantels. xx + 314pp. 7⅞ x 10¾.
21743-4 Paperbound $3.50

EARLY AMERICAN ROOMS: 1650-1858, edited by Russell Hawes Kettell. Tour of 12 rooms, each representative of a different era in American history and each furnished, decorated, designed and occupied in the style of the era. 72 plans and elevations, 8-page color section, etc., show fabrics, wall papers, arrangements, etc. Full descriptive text. xvii + 200pp. of text. 8⅜ x 11¼.
21633-0 Paperbound $4.00

THE FITZWILLIAM VIRGINAL BOOK, edited by J. Fuller Maitland and W. B. Squire. Full modern printing of famous early 17th-century ms. volume of 300 works by Morley, Byrd, Bull, Gibbons, etc. For piano or other modern keyboard instrument; easy to read format. xxxvi + 938pp. 8⅜ x 11.
21068-5, 21069-3 Two volumes, Paperbound $8.00

HARPSICHORD MUSIC, Johann Sebastian Bach. Bach Gesellschaft edition. A rich selection of Bach's masterpieces for the harpsichord: the six English Suites, six French Suites, the six Partitas (Clavierübung part I), the Goldberg Variations (Clavierübung part IV), the fifteen Two-Part Inventions and the fifteen Three-Part Sinfonias. Clearly reproduced on large sheets with ample margins; eminently playable. vi + 312pp. 8⅛ x 11.
22360-4 Paperbound $5.00

THE MUSIC OF BACH: AN INTRODUCTION, Charles Sanford Terry. A fine, nontechnical introduction to Bach's music, both instrumental and vocal. Covers organ music, chamber music, passion music, other types. Analyzes themes, developments, innovations. x + 114pp.
21075-8 Paperbound $1.25

BEETHOVEN AND HIS NINE SYMPHONIES, Sir George Grove. Noted British musicologist provides best history, analysis, commentary on symphonies. Very thorough, rigorously accurate; necessary to both advanced student and amateur music lover. 436 musical passages. vii + 407 pp.
20334-4 Paperbound $2.25

JOHANN SEBASTIAN BACH, Philipp Spitta. One of the great classics of musicology, this definitive analysis of Bach's music (and life) has never been surpassed. Lucid, nontechnical analyses of hundreds of pieces (30 pages devoted to St. Matthew Passion, 26 to B Minor Mass). Also includes major analysis of 18th-century music. 450 musical examples. 40-page musical supplement. Total of xx + 1799pp.
(EUK) 22278-0, 22279-9 Two volumes, Clothbound $15.00

MOZART AND HIS PIANO CONCERTOS, Cuthbert Girdlestone. The only full-length study of an important area of Mozart's creativity. Provides detailed analyses of all 23 concertos, traces inspirational sources. 417 musical examples. Second edition. 509pp. (USO) 21271-8 Paperbound $2.50

THE PERFECT WAGNERITE: A COMMENTARY ON THE NIBLUNG'S RING, George Bernard Shaw. Brilliant and still relevant criticism in remarkable essays on Wagner's Ring cycle, Shaw's ideas on political and social ideology behind the plots, role of Leitmotifs, vocal requisites, etc. Prefaces. xxi + 136pp.
21707-8 Paperbound $1.50

DON GIOVANNI, W. A. Mozart. Complete libretto, modern English translation; biographies of composer and librettist; accounts of early performances and critical reaction. Lavishly illustrated. All the material you need to understand and appreciate this great work. Dover Opera Guide and Libretto Series; translated and introduced by Ellen Bleiler. 92 illustrations. 209pp.
21134-7 Paperbound $1.50

HIGH FIDELITY SYSTEMS: A LAYMAN'S GUIDE, Roy F. Allison. All the basic information you need for setting up your own audio system: high fidelity and stereo record players, tape records, F.M. Connections, adjusting tone arm, cartridge, checking needle alignment, positioning speakers, phasing speakers, adjusting hums, trouble-shooting, maintenance, and similar topics. Enlarged 1965 edition. More than 50 charts, diagrams, photos. iv + 91pp. 21514-8 Paperbound $1.25

REPRODUCTION OF SOUND, Edgar Villchur. Thorough coverage for laymen of high fidelity systems, reproducing systems in general, needles, amplifiers, preamps, loudspeakers, feedback, explaining physical background. "A rare talent for making technicalities vividly comprehensible," R. Darrell, *High Fidelity*. 69 figures. iv + 92pp. 21515-6 Paperbound $1.00

HEAR ME TALKIN' TO YA: THE STORY OF JAZZ AS TOLD BY THE MEN WHO MADE IT, Nat Shapiro and Nat Hentoff. Louis Armstrong, Fats Waller, Jo Jones, Clarence Williams, Billy Holiday, Duke Ellington, Jelly Roll Morton and dozens of other jazz greats tell how it was in Chicago's South Side, New Orleans, depression Harlem and the modern West Coast as jazz was born and grew. xvi + 429pp.
21726-4 Paperbound $2.00

FABLES OF AESOP, translated by Sir Roger L'Estrange. A reproduction of the very rare 1931 Paris edition; a selection of the most interesting fables, together with 50 imaginative drawings by Alexander Calder. v + 128pp. 6½x9¼.
21780-9 Paperbound $1.25

AGAINST THE GRAIN (A REBOURS), Joris K. Huysmans. Filled with weird images, evidences of a bizarre imagination, exotic experiments with hallucinatory drugs, rich tastes and smells and the diversions of its sybarite hero Duc Jean des Esseintes, this classic novel pushed 19th-century literary decadence to its limits. Full unabridged edition. Do not confuse this with abridged editions generally sold. Introduction by Havelock Ellis. xlix + 206pp.   22190-3 Paperbound $2.00

VARIORUM SHAKESPEARE: HAMLET. Edited by Horace H. Furness; a landmark of American scholarship. Exhaustive footnotes and appendices treat all doubtful words and phrases, as well as suggested critical emendations throughout the play's history. First volume contains editor's own text, collated with all Quartos and Folios. Second volume contains full first Quarto, translations of Shakespeare's sources (Belleforest, and Saxo Grammaticus), Der Bestrafte Brudermord, and many essays on critical and historical points of interest by major authorities of past and present. Includes details of staging and costuming over the years. By far the best edition available for serious students of Shakespeare. Total of xx + 905pp.
21004-9, 21005-7, 2 volumes, Paperbound $5.25

A LIFE OF WILLIAM SHAKESPEARE, Sir Sidney Lee. This is the standard life of Shakespeare, summarizing everything known about Shakespeare and his plays. Incredibly rich in material, broad in coverage, clear and judicious, it has served thousands as the best introduction to Shakespeare. 1931 edition. 9 plates. xxix + 792pp.   (USO) 21967-4 Paperbound $3.75

MASTERS OF THE DRAMA, John Gassner. Most comprehensive history of the drama in print, covering every tradition from Greeks to modern Europe and America, including India, Far East, etc. Covers more than 800 dramatists, 2000 plays, with biographical material, plot summaries, theatre history, criticism, etc. "Best of its kind in English," *New Republic*. 77 illustrations. xxii + 890pp.
20100-7 Clothbound $7.50

THE EVOLUTION OF THE ENGLISH LANGUAGE, George McKnight. The growth of English, from the 14th century to the present. Unusual, non-technical account presents basic information in very interesting form: sound shifts, change in grammar and syntax, vocabulary growth, similar topics. Abundantly illustrated with quotations. Formerly *Modern English in the Making*. xii + 590pp.
21932-1 Paperbound $3.50

AN ETYMOLOGICAL DICTIONARY OF MODERN ENGLISH, Ernest Weekley. Fullest, richest work of its sort, by foremost British lexicographer. Detailed word histories, including many colloquial and archaic words; extensive quotations. Do not confuse this with the Concise Etymological Dictionary, which is much abridged. Total of xxvii + 830pp. 6½ x 9¼.
21873-2, 21874-0 Two volumes, Paperbound $5.50

FLATLAND: A ROMANCE OF MANY DIMENSIONS, E. A. Abbott. Classic of science-fiction explores ramifications of life in a two-dimensional world, and what happens when a three-dimensional being intrudes. Amusing reading, but also useful as introduction to thought about hyperspace. Introduction by Banesh Hoffmann. 16 illustrations. xx + 103pp.   20001-9 Paperbound $1.00

POEMS OF ANNE BRADSTREET, edited with an introduction by Robert Hutchinson. A new selection of poems by America's first poet and perhaps the first significant woman poet in the English language. 48 poems display her development in works of considerable variety—love poems, domestic poems, religious meditations, formal elegies, "quaternions," etc. Notes, bibliography. viii + 222pp.
22160-1 Paperbound $2.00

THREE GOTHIC NOVELS: THE CASTLE OF OTRANTO BY HORACE WALPOLE; VATHEK BY WILLIAM BECKFORD; THE VAMPYRE BY JOHN POLIDORI, WITH FRAGMENT OF A NOVEL BY LORD BYRON, edited by E. F. Bleiler. The first Gothic novel, by Walpole; the finest Oriental tale in English, by Beckford; powerful Romantic supernatural story in versions by Polidori and Byron. All extremely important in history of literature; all still exciting, packed with supernatural thrills, ghosts, haunted castles, magic, etc. xl + 291pp.
21232-7 Paperbound $2.00

THE BEST TALES OF HOFFMANN, E. T. A. Hoffmann. 10 of Hoffmann's most important stories, in modern re-editings of standard translations: Nutcracker and the King of Mice, Signor Formica, Automata, The Sandman, Rath Krespel, The Golden Flowerpot, Master Martin the Cooper, The Mines of Falun, The King's Betrothed, A New Year's Eve Adventure. 7 illustrations by Hoffmann. Edited by E. F. Bleiler. xxxix + 419pp.
21793-0 Paperbound $2.25

GHOST AND HORROR STORIES OF AMBROSE BIERCE, Ambrose Bierce. 23 strikingly modern stories of the horrors latent in the human mind: The Eyes of the Panther, The Damned Thing, An Occurrence at Owl Creek Bridge, An Inhabitant of Carcosa, etc., plus the dream-essay, Visions of the Night. Edited by E. F. Bleiler. xxii + 199pp.
20767-6 Paperbound $1.50

BEST GHOST STORIES OF J. S. LeFANU, J. Sheridan LeFanu. Finest stories by Victorian master often considered greatest supernatural writer of all. Carmilla, Green Tea, The Haunted Baronet, The Familiar, and 12 others. Most never before available in the U. S. A. Edited by E. F. Bleiler. 8 illustrations from Victorian publications. xvii + 467pp.
20415-4 Paperbound $2.50

THE TIME STREAM, THE GREATEST ADVENTURE, AND THE PURPLE SAPPHIRE—THREE SCIENCE FICTION NOVELS, John Taine (Eric Temple Bell). Great American mathematician was also foremost science fiction novelist of the 1920's. *The Time Stream,* one of all-time classics, uses concepts of circular time; *The Greatest Adventure,* incredibly ancient biological experiments from Antarctica threaten to escape; The *Purple Sapphire,* superscience, lost races in Central Tibet, survivors of the Great Race. 4 illustrations by Frank R. Paul. v + 532pp.
21180-0 Paperbound $2.50

SEVEN SCIENCE FICTION NOVELS, H. G. Wells. The standard collection of the great novels. Complete, unabridged. *First Men in the Moon, Island of Dr. Moreau, War of the Worlds, Food of the Gods, Invisible Man, Time Machine, In the Days of the Comet.* Not only science fiction fans, but every educated person owes it to himself to read these novels. 1015pp.
20264-X Clothbound $5.00

LAST AND FIRST MEN AND STAR MAKER, TWO SCIENCE FICTION NOVELS, Olaf Stapledon. Greatest future histories in science fiction. In the first, human intelligence is the "hero," through strange paths of evolution, interplanetary invasions, incredible technologies, near extinctions and reemergences. Star Maker describes the quest of a band of star rovers for intelligence itself, through time and space: weird inhuman civilizations, crustacean minds, symbiotic worlds, etc. Complete, unabridged. v + 438pp. 21962-3 Paperbound $2.00

THREE PROPHETIC NOVELS, H. G. WELLS. Stages of a consistently planned future for mankind. *When the Sleeper Wakes,* and *A Story of the Days to Come,* anticipate *Brave New World* and *1984,* in the 21st Century; *The Time Machine,* only complete version in print, shows farther future and the end of mankind. All show Wells's greatest gifts as storyteller and novelist. Edited by E. F. Bleiler. x + 335pp. (USO) 20605-X Paperbound $2.00

THE DEVIL'S DICTIONARY, Ambrose Bierce. America's own Oscar Wilde—Ambrose Bierce—offers his barbed iconoclastic wisdom in over 1,000 definitions hailed by H. L. Mencken as "some of the most gorgeous witticisms in the English language." 145pp. 20487-1 Paperbound $1.25

MAX AND MORITZ, Wilhelm Busch. Great children's classic, father of comic strip, of two bad boys, Max and Moritz. Also Ker and Plunk (Plisch und Plumm), Cat and Mouse, Deceitful Henry, Ice-Peter, The Boy and the Pipe, and five other pieces. Original German, with English translation. Edited by H. Arthur Klein; translations by various hands and H. Arthur Klein. vi + 216pp. 20181-3 Paperbound $1.50

PIGS IS PIGS AND OTHER FAVORITES, Ellis Parker Butler. The title story is one of the best humor short stories, as Mike Flannery obfuscates biology and English. Also included, That Pup of Murchison's, The Great American Pie Company, and Perkins of Portland. 14 illustrations. v + 109pp. 21532-6 Paperbound $1.00

THE PETERKIN PAPERS, Lucretia P. Hale. It takes genius to be as stupidly mad as the Peterkins, as they decide to become wise, celebrate the "Fourth," keep a cow, and otherwise strain the resources of the Lady from Philadelphia. Basic book of American humor. 153 illustrations. 219pp. 20794-3 Paperbound $1.25

PERRAULT'S FAIRY TALES, translated by A. E. Johnson and S. R. Littlewood, with 34 full-page illustrations by Gustave Doré. All the original Perrault stories—Cinderella, Sleeping Beauty, Bluebeard, Little Red Riding Hood, Puss in Boots, Tom Thumb, etc.—with their witty verse morals and the magnificent illustrations of Doré. One of the five or six great books of European fairy tales. viii + 117pp. 8⅛ x 11. 22311-6 Paperbound $2.00

OLD HUNGARIAN FAIRY TALES, Baroness Orczy. Favorites translated and adapted by author of the *Scarlet Pimpernel.* Eight fairy tales include "The Suitors of Princess Fire-Fly," "The Twin Hunchbacks," "Mr. Cuttlefish's Love Story," and "The Enchanted Cat." This little volume of magic and adventure will captivate children as it has for generations. 90 drawings by Montagu Barstow. 96pp. (USO) 22293-4 Paperbound $1.95

THE RED FAIRY BOOK, Andrew Lang. Lang's color fairy books have long been children's favorites. This volume includes Rapunzel, Jack and the Bean-stalk and 35 other stories, familiar and unfamiliar. 4 plates, 93 illustrations x + 367pp.
21673-X Paperbound $1.95

THE BLUE FAIRY BOOK, Andrew Lang. Lang's tales come from all countries and all times. Here are 37 tales from Grimm, the Arabian Nights, Greek Mythology, and other fascinating sources. 8 plates, 130 illustrations. xi + 390pp.
21437-0 Paperbound $1.95

HOUSEHOLD STORIES BY THE BROTHERS GRIMM. Classic English-language edition of the well-known tales — Rumpelstiltskin, Snow White, Hansel and Gretel, The Twelve Brothers, Faithful John, Rapunzel, Tom Thumb (52 stories in all). Translated into simple, straightforward English by Lucy Crane. Ornamented with headpieces, vignettes, elaborate decorative initials and a dozen full-page illustrations by Walter Crane. x + 269pp.
21080-4 Paperbound $1.75

THE MERRY ADVENTURES OF ROBIN HOOD, Howard Pyle. The finest modern versions of the traditional ballads and tales about the great English outlaw. Howard Pyle's complete prose version, with every word, every illustration of the first edition. Do not confuse this facsimile of the original (1883) with modern editions that change text or illustrations. 23 plates plus many page decorations. xxii + 296pp.
22043-5 Paperbound $2.00

THE STORY OF KING ARTHUR AND HIS KNIGHTS, Howard Pyle. The finest children's version of the life of King Arthur; brilliantly retold by Pyle, with 48 of his most imaginative illustrations. xviii + 313pp. 6⅛ x 9¼.
21445-1 Paperbound $2.00

THE WONDERFUL WIZARD OF OZ, L. Frank Baum. America's finest children's book in facsimile of first edition with all Denslow illustrations in full color. The edition a child should have. Introduction by Martin Gardner. 23 color plates, scores of drawings. iv + 267pp.
20691-2 Paperbound $1.95

THE MARVELOUS LAND OF OZ, L. Frank Baum. The second Oz book, every bit as imaginative as the Wizard. The hero is a boy named Tip, but the Scarecrow and the Tin Woodman are back, as is the Oz magic. 16 color plates, 120 drawings by John R. Neill. 287pp.
20692-0 Paperbound $1.75

THE MAGICAL MONARCH OF MO, L. Frank Baum. Remarkable adventures in a land even stranger than Oz. The best of Baum's books not in the Oz series. 15 color plates and dozens of drawings by Frank Verbeck. xviii + 237pp.
21892-9 Paperbound $2.00

THE BAD CHILD'S BOOK OF BEASTS, MORE BEASTS FOR WORSE CHILDREN, A MORAL ALPHABET, Hilaire Belloc. Three complete humor classics in one volume. Be kind to the frog, and do not call him names . . . and 28 other whimsical animals. Familiar favorites and some not so well known. Illustrated by Basil Blackwell. 156pp.
(USO) 20749-8 Paperbound $1.25

EAST O' THE SUN AND WEST O' THE MOON, George W. Dasent. Considered the best of all translations of these Norwegian folk tales, this collection has been enjoyed by generations of children (and folklorists too). Includes True and Untrue, Why the Sea is Salt, East O' the Sun and West O' the Moon, Why the Bear is Stumpy-Tailed, Boots and the Troll, The Cock and the Hen, Rich Peter the Pedlar, and 52 more. The only edition with all 59 tales. 77 illustrations by Erik Werenskiold and Theodor Kittelsen. xv + 418pp.   22521-6 Paperbound $3.00

GOOPS AND HOW TO BE THEM, Gelett Burgess. Classic of tongue-in-cheek humor, masquerading as etiquette book. 87 verses, twice as many cartoons, show mischievous Goops as they demonstrate to children virtues of table manners, neatness, courtesy, etc. Favorite for generations. viii + 88pp. $6\frac{1}{2}$ x $9\frac{1}{4}$.
                                             22233-0 Paperbound $1.25

ALICE'S ADVENTURES UNDER GROUND, Lewis Carroll. The first version, quite different from the final Alice in Wonderland, printed out by Carroll himself with his own illustrations. Complete facsimile of the "million dollar" manuscript Carroll gave to Alice Liddell in 1864. Introduction by Martin Gardner. viii + 96pp. Title and dedication pages in color.   21482-6 Paperbound $1.00

THE BROWNIES, THEIR BOOK, Palmer Cox. Small as mice, cunning as foxes, exuberant and full of mischief, the Brownies go to the zoo, toy shop, seashore, circus, etc., in 24 verse adventures and 266 illustrations. Long a favorite, since their first appearance in St. Nicholas Magazine. xi + 144pp. $6\frac{5}{8}$ x $9\frac{1}{4}$.
                                             21265-3 Paperbound $1.50

SONGS OF CHILDHOOD, Walter De La Mare. Published (under the pseudonym Walter Ramal) when De La Mare was only 29, this charming collection has long been a favorite children's book. A facsimile of the first edition in paper, the 47 poems capture the simplicity of the nursery rhyme and the ballad, including such lyrics as I Met Eve, Tartary, The Silver Penny. vii + 106pp.   21972-0 Paperbound $1.25

THE COMPLETE NONSENSE OF EDWARD LEAR, Edward Lear. The finest 19th-century humorist-cartoonist in full: all nonsense limericks, zany alphabets, Owl and Pussycat, songs, nonsense botany, and more than 500 illustrations by Lear himself. Edited by Holbrook Jackson. xxix + 287pp.   (USO) 20167-8 Paperbound $1.75

BILLY WHISKERS: THE AUTOBIOGRAPHY OF A GOAT, Frances Trego Montgomery. A favorite of children since the early 20th century, here are the escapades of that rambunctious, irresistible and mischievous goat—Billy Whiskers. Much in the spirit of Peck's Bad Boy, this is a book that children never tire of reading or hearing. All the original familiar illustrations by W. H. Fry are included: 6 color plates, 18 black and white drawings. 159pp.   22345-0 Paperbound $2.00

MOTHER GOOSE MELODIES. Faithful republication of the fabulously rare Munroe and Francis "copyright 1833" Boston edition—the most important Mother Goose collection, usually referred to as the "original." Familiar rhymes plus many rare ones, with wonderful old woodcut illustrations. Edited by E. F. Bleiler. 128pp. $4\frac{1}{2}$ x $6\frac{3}{8}$.   22577-1 Paperbound $1.25

TWO LITTLE SAVAGES; BEING THE ADVENTURES OF TWO BOYS WHO LIVED AS INDIANS AND WHAT THEY LEARNED, Ernest Thompson Seton. Great classic of nature and boyhood provides a vast range of woodlore in most palatable form, a genuinely entertaining story. Two farm boys build a teepee in woods and live in it for a month, working out Indian solutions to living problems, star lore, birds and animals, plants, etc. 293 illustrations. vii + 286pp.

20985-7 Paperbound $1.95

PETER PIPER'S PRACTICAL PRINCIPLES OF PLAIN & PERFECT PRONUNCIATION. Alliterative jingles and tongue-twisters of surprising charm, that made their first appearance in America about 1830. Republished in full with the spirited woodcut illustrations from this earliest American edition. 32pp. 4½ x 6⅜.

22560-7 Paperbound $1.00

SCIENCE EXPERIMENTS AND AMUSEMENTS FOR CHILDREN, Charles Vivian. 73 easy experiments, requiring only materials found at home or easily available, such as candles, coins, steel wool, etc.; illustrate basic phenomena like vacuum, simple chemical reaction, etc. All safe. Modern, well-planned. Formerly *Science Games for Children*. 102 photos, numerous drawings. 96pp. 6⅛ x 9¼.

21856-2 Paperbound $1.25

AN INTRODUCTION TO CHESS MOVES AND TACTICS SIMPLY EXPLAINED, Leonard Barden. Informal intermediate introduction, quite strong in explaining reasons for moves. Covers basic material, tactics, important openings, traps, positional play in middle game, end game. Attempts to isolate patterns and recurrent configurations. Formerly *Chess*. 58 figures. 102pp. (USO) 21210-6 Paperbound $1.25

LASKER'S MANUAL OF CHESS, Dr. Emanuel Lasker. Lasker was not only one of the five great World Champions, he was also one of the ablest expositors, theorists, and analysts. In many ways, his Manual, permeated with his philosophy of battle, filled with keen insights, is one of the greatest works ever written on chess. Filled with analyzed games by the great players. A single-volume library that will profit almost any chess player, beginner or master. 308 diagrams. xli x 349pp.

20640-8 Paperbound $2.50

THE MASTER BOOK OF MATHEMATICAL RECREATIONS, Fred Schuh. In opinion of many the finest work ever prepared on mathematical puzzles, stunts, recreations; exhaustively thorough explanations of mathematics involved, analysis of effects, citation of puzzles and games. Mathematics involved is elementary. Translated by F. Göbel. 194 figures. xxiv + 430pp. 22134-2 Paperbound $3.00

MATHEMATICS, MAGIC AND MYSTERY, Martin Gardner. Puzzle editor for Scientific American explains mathematics behind various mystifying tricks: card tricks, stage "mind reading," coin and match tricks, counting out games, geometric dissections, etc. Probability sets, theory of numbers clearly explained. Also provides more than 400 tricks, guaranteed to work, that you can do. 135 illustrations. xii + 176pp.

20338-2 Paperbound $1.50

MATHEMATICAL PUZZLES FOR BEGINNERS AND ENTHUSIASTS, Geoffrey Mott-Smith. 189 puzzles from easy to difficult—involving arithmetic, logic, algebra, properties of digits, probability, etc.—for enjoyment and mental stimulus. Explanation of mathematical principles behind the puzzles. 135 illustrations. viii + 248pp.
20198-8 Paperbound $1.25

PAPER FOLDING FOR BEGINNERS, William D. Murray and Francis J. Rigney. Easiest book on the market, clearest instructions on making interesting, beautiful origami. Sail boats, cups, roosters, frogs that move legs, bonbon boxes, standing birds, etc. 40 projects; more than 275 diagrams and photographs. 94pp.
20713-7 Paperbound $1.00

TRICKS AND GAMES ON THE POOL TABLE, Fred Herrmann. 79 tricks and games— some solitaires, some for two or more players, some competitive games—to entertain you between formal games. Mystifying shots and throws, unusual caroms, tricks involving such props as cork, coins, a hat, etc. Formerly *Fun on the Pool Table*. 77 figures. 95pp.
21814-7 Paperbound $1.00

HAND SHADOWS TO BE THROWN UPON THE WALL: A SERIES OF NOVEL AND AMUSING FIGURES FORMED BY THE HAND, Henry Bursill. Delightful picturebook from great-grandfather's day shows how to make 18 different hand shadows: a bird that flies, duck that quacks, dog that wags his tail, camel, goose, deer, boy, turtle, etc. Only book of its sort. vi + 33pp. 6½ x 9¼. 21779-5 Paperbound $1.00

WHITTLING AND WOODCARVING, E. J. Tangerman. 18th printing of best book on market. "If you can cut a potato you can carve" toys and puzzles, chains, chessmen, caricatures, masks, frames, woodcut blocks, surface patterns, much more. Information on tools, woods, techniques. Also goes into serious wood sculpture from Middle Ages to present, East and West. 464 photos, figures. x + 293pp.
20965-2 Paperbound $2.00

HISTORY OF PHILOSOPHY, Julián Marias. Possibly the clearest, most easily followed, best planned, most useful one-volume history of philosophy on the market; neither skimpy nor overfull. Full details on system of every major philosopher and dozens of less important thinkers from pre-Socratics up to Existentialism and later. Strong on many European figures usually omitted. Has gone through dozens of editions in Europe. 1966 edition, translated by Stanley Appelbaum and Clarence Strowbridge. xviii + 505pp. 21739-6 Paperbound $2.75

YOGA: A SCIENTIFIC EVALUATION, Kovoor T. Behanan. Scientific but non-technical study of physiological results of yoga exercises; done under auspices of Yale U. Relations to Indian thought, to psychoanalysis, etc. 16 photos. xxiii + 270pp.
20505-3 Paperbound $2.50

Prices subject to change without notice.
Available at your book dealer or write for free catalogue to Dept. GI, Dover Publications, Inc., 180 Varick St., N. Y., N. Y. 10014. Dover publishes more than 150 books each year on science, elementary and advanced mathematics, biology, music, art, literary history, social sciences and other areas.

MATHEMATICAL PUZZLES FOR BEGINNERS AND ENTHUSIASTS, Geoffrey Mott-Smith. 189 puzzles from easy to difficult—involving arithmetic, logic, algebra, properties of digits, probability, etc.—for enjoyment and mental stimulus. Explanation of mathematical principles behind the puzzles. 135 illustrations. viii + 248pp.
20198-8 Paperbound $1.25

PAPER FOLDING FOR BEGINNERS, William D. Murray and Francis J. Rigney. Easiest book on the market, clearest instructions on making interesting, beautiful origami. Sail boats, cups, roosters, frogs that move legs, bonbon boxes, standing birds, etc. 40 projects; more than 275 diagrams and photographs. 94pp.
20713-7 Paperbound $1.00

TRICKS AND GAMES ON THE POOL TABLE, Fred Herrmann. 79 tricks and games—some solitaires, some for two or more players, some competitive games—to entertain you between formal games. Mystifying shots and throws, unusual caroms, tricks involving such props as cork, coins, a hat, etc. Formerly *Fun on the Pool Table*. 77 figures. 95pp.
21814-7 Paperbound $1.00

HAND SHADOWS TO BE THROWN UPON THE WALL: A SERIES OF NOVEL AND AMUSING FIGURES FORMED BY THE HAND, Henry Bursill. Delightful picturebook from great-grandfather's day shows how to make 18 different hand shadows: a bird that flies, duck that quacks, dog that wags his tail, camel, goose, deer, boy, turtle, etc. Only book of its sort. vi + 33pp. 6½ x 9¼. 21779-5 Paperbound $1.00

WHITTLING AND WOODCARVING, E. J. Tangerman. 18th printing of best book on market. "If you can cut a potato you can carve" toys and puzzles, chains, chessmen, caricatures, masks, frames, woodcut blocks, surface patterns, much more. Information on tools, woods, techniques. Also goes into serious wood sculpture from Middle Ages to present, East and West. 464 photos, figures. x + 293pp.
20965-2 Paperbound $2.00

HISTORY OF PHILOSOPHY, Julián Marias. Possibly the clearest, most easily followed, best planned, most useful one-volume history of philosophy on the market; neither skimpy nor overfull. Full details on system of every major philosopher and dozens of less important thinkers from pre-Socratics up to Existentialism and later. Strong on many European figures usually omitted. Has gone through dozens of editions in Europe. 1966 edition, translated by Stanley Appelbaum and Clarence Strowbridge. xviii + 505pp.
21739-6 Paperbound $2.75

YOGA: A SCIENTIFIC EVALUATION, Kovoor T. Behanan. Scientific but non-technical study of physiological results of yoga exercises; done under auspices of Yale U. Relations to Indian thought, to psychoanalysis, etc. 16 photos. xxiii + 270pp.
20505-3 Paperbound $2.50

Prices subject to change without notice.
Available at your book dealer or write for free catalogue to Dept. GI, Dover Publications, Inc., 180 Varick St., N. Y., N. Y. 10014. Dover publishes more than 150 books each year on science, elementary and advanced mathematics, biology, music, art, literary history, social sciences and other areas.

(continued from front flap)

FRESHWATER MICROSCOPY, W. J. Garnett. (60790-9) Clothbound $5.95

ANIMAL GEOGRAPHY, Wilma George. (21294-7) Clothbound $5.00

STUDIES ON THE STRUCTURE AND DEVELOPMENT OF VERTEBRATES, E. S. Goodrich. (60449-7, 60450-0) Two-volume set $7.00

EXTINCT AND VANISHING BIRDS OF THE WORLD, James C. Greenway, Jr. (21869-4) $3.50

BIOLOGY EXPERIMENTS FOR CHILDREN, Ethel Hanauer. (22032-X) $1.25

FRUIT KEY AND TWIG KEY TO TREES AND SHRUBS, William M. Harlow. (20511-8) $1.35

TREES OF THE EASTERN AND CENTRAL UNITED STATES AND CANADA, William M. Harlow. (20395-6) $1.50

GUIDE TO SOUTHERN TREES, Ellwood S. Harrar and J. George Harrar. (20945-8) $3.00

THE HANDBOOK OF PLANT AND FLORAL ORNAMENT FROM EARLY HERBALS, Richard G. Hatton. (20649-1) $4.00

THE PSYCHOLOGY AND BEHAVOIR OF ANIMALS IN ZOOS AND CIRCUSES, H. Hediger. (62218-5) $2.00

WILD ANIMALS IN CAPTIVITY: AN OUTLINE OF THE BIOLOGY OF ZOOLOGICAL GARDENS, H. Hediger. (21260-2) $2.50

BIG FLEAS HAVE LITTLE FLEAS, OR, WHO'S WHO AMONG THE PROTOZOA, Robert Hegner. (22040-0) $2.50

TREATISE ON PHYSIOLOGICAL OPTICS, Herman Helmholtz. (60015-7, 60016-5) Two-volume set, Clothbound $20.00

THE MOTH BOOK, W. J. Holland. (21948-8) $5.00

THE STRUGGLE FOR EXISTENCE

G. F. GAUSE

From Russian microbiologist G. F. Gause, recipient of the coveted Stalin prize, here is the work that broke ground for all subsequent research in the biomathemathics of populations and competition between them. In it, the reader will find not only a lucid introduction to the tools of mathematical biology, but a detailed account of Gause's own extensive and important researches with microbial populations as well.

Gause begins with an exploration of the struggle for existence in nature (with special reference to botanical data) and goes on to summarize the important theoretical and experimental work that preceded his own (that of Pearl, Lotka, Reed, Volterra, and others). Following a discussion of the tools of mathematical biology, formulas are derived which make possible the quantitative measurement of such basic concepts as potential population increase, population saturation, environmental resistance, and the intensity of the struggle for existence. Gause then proceeds to report in depth on his own experimental work with unicellular populations. His experiments on the mechanisms of indirect competition (for food and space) with two species of yeast, similar experiments with Protozoan populations, and researchers in direct competition (involving the destruction of one species by another) performed among one Bacillus and two Protozoan populations, are reported in detail. His important conclusions, that the periodic expansions and contractions of populations are dependent upon the introduction of new variables and are not an inherent property of the predator-prey relationship, is clearly presented.

Gause's work was essential in giving modern science its grasp of the complexities of population competition; it provides a microcosm of the processes at work on a larger scale throughout the biosphere, and gives the reader the means for quantitative evaluation of them. Easily understood by anyone who has an acquaintance with higher mathematics, this book provides the background necessary to every modern student of population dynamics.

Unaltered, unabridged republication of original (1934) edition. Foreword by Raymond Pearl. Author's Preface. 2 appendices. Bibliography. Index. 40 figures; 20 tables. ix + 163pp. 5⅜ x 8½.

22697-2 Paperbound

A DOVER EDITION DESIGNED FOR YEARS OF USE!

We have made every effort to make this the best book possible. Our paper is opaque, with minimal show-through; it will not discolor or become brittle with age. Pages are sewn in signatures, in the method traditionally used for the best books, and will not drop out, as often happens with paperbacks held together with glue. Books open flat for easy reference. The binding will not crack or split. This is a permanent book.